BALES FARMS
Cookbook

By Aliceson Bales

Bales Farms Cookbook

© 2021 by Aliceson Bales

Printed in the United States of America

1 2 3 4 5 6 7 8 9 10
ISBN 978-0-578-93831-8

Contents

ESTD.

1882

BALES FARMS

MOSHEIM, TN.

Foreword by Dolly Parton

Two things I consider myself an expert in is knowing good people and good food. In Bales Farms Cookbook both things get a big gold star from me. Aliceson is one of the most beautiful women, inside and out, that I have ever met. . . and what a cook.

Barry is as great a musician as Aliceson is a cook. He is equally as great a human being. Barry has played bass on some of my best records. He dresses like a farmer, he looks like a farmer, he acts like a farmer. . . down to the big wad of Burley Tobacco in his jaw. He even has his own personal spit can in the studio!

I'm not surprised that Barry and Aliceson have their own farm, growing their own food and sharing it with all of us. I personally am going to cook every recipe in the cookbook. Not only do I know it will be great, it will also make me feel close to two people that I love, respect and admire. I will think of them while I'm cooking and eating it and sharing it with others that I love. I wish you much luck with this cookbook. I'm sure that I'm going to enjoy it and I think the rest of the world will, too.

Love,

Aliceson's story

I love to eat. I love all the aspects of food - the flavors, the textures, the feelings from the food and company we keep while we eat and even the time I spend in the kitchen preparing food.

I was blessed to grow up around great cooks. My mom is a fantastic cook and her baking skills?? No one can beat her. Her sourdough bread is FAMOUS. Truly famous. People beg her for her bread and she bakes 6 loaves each week. And when I was little she would make these unbelievably beautiful and mouthwatering cheese braids for Christmas gifts. They were absolutely breathtaking. If she had social media she would've been INSTAFAMOUS for sure. But it was the 80s. And when my sister or I had slumber parties she made homemade donuts for us for breakfast and donut holes. (The holes are always the best, people.) After church on Sundays she'd whip up homemade crepes and fill them with fresh strawberries. Can you imagine??

And my paternal grandmother, Mamaw (Grace), was also a phenomenal cook. Her meringue was at least a mile high on her pies (which she cut into 4 pieces. FOUR. It's a wonder any of us could fit through the door.). How she achieved that in a hot kitchen on a summer afternoon with no air conditioning is beyond me (or why she'd even feel like cooking in those conditions). She made homemade potato chips and the most amazing fried potatoes you've ever ever put in your mouth. Her kitchen table was sturdy - it had to be - because it held mounds of food. There was no one pot and done meal on her table. Oh no. She would have 2 meats and more vegetables and fruit than you could count.

So I was blessed with great cooks who had a heart for teaching a snippy, lazy, immature knuckle-head like me. They taught me how to make biscuits from scratch, and when I messed the batter up how to fix it without throwing it away and starting over (because we didn't throw things away back then. We fixed them and used them). They taught me step by step. And while I don't have the baking skills my mom has or Mamaw's exact fried potatoes, I appreciate the time they took and the love they showed. And I am so thankful I can still call my mom when I mess up and she'll talk me through my mess to get me back on track.

So fast forward to adulthood. Barry and I have been married now over 20 years. Over 2 decades! I've now known my husband for over half my life. Sometimes that fact blows my mind. When we dated our time together was so limited (he was a traveling musician and I was a full time practicing Physical Therapist) so when we actually got to see each other we didn't do things like go to movies or concerts. We spent time together on his farm or in my apartment and we cooked and ate and enjoyed the company. After we got married and I moved to his farm we were so far out it just never made sense to eat out much. We've always cooked our food and we've always enjoyed good food. One of our favorite nights of the week was Friday night - we'd make fajitas and mango margaritas. Those nights are great memories to me and when I think about our early marriage that's what I remember. We'd enjoy our drinks and talk and laugh about the day while chopping, dicing, sautéing and sipping. Such great moments in that old kitchen in the old house. Things haven't changes THAT much since then, at least with food. We've gotten bolder and more adventurous in our cooking and tasting. We're more mature in food and drink choices. We have learned not to hurry in the kitchen. And we've realized great food makes fond memories when shared in a relaxed, unhurried atmosphere among people you love.

Since I love food I have learned to love the preparation. It does slightly boggle my mind to hear people say they just hate to cook. I get that sometimes at the end of a busy day after work and soccer and dancing and homework it's just not convenient to spend an hour or two in the kitchen but you can still whip up an amazing meal in about 20 minutes. I encourage you, if you're in that boat, to turn off the tv, put down the phone, turn on some music that you love and lose yourself in the joy that is found in loving your people through food. Spend time unwinding with the food and

the music and then sit down together at the table, talking, laughing and detailing your day. Live an unhurried life, if even for an hour a day.

And I believe if you connect yourself to the food in some way it helps tremendously. Growing a garden or even having a planter by the door filled with herbs or a tomato plant or two - that connects you with the ingredients. (And it helps get littles involved, too.) I always say using fresh herbs in a meal or a drink transcends the finished product. And fresh basil out of the garden?? Oh my heavens! It makes the best pesto (which you can make in 30 seconds) and an even greater cucumber-basil martini. So get connected. Learn the art of eating what you pick from the garden or the planter by the door. Anecdotally I'll say we grow spring peas. If you've only had peas from a can I apologize for your sheltered existence because when you pick and shell peas you won't even make it to the kitchen with them. We eat them before we even get in the door. They're so fresh and sweet and plump and pop in your mouth!! Oh. Oh. Oh.

My hope for this book is that as you go through it you'll find some recipes to try. Some are pretty simple! In fact I'd say most are simple with ingredients you have already in your pantry. Most take less than an hour total to prepare. There are a few here and there that are better suited for the weekend for sure (pulled pork barbecue for instance) but for the most part you can have dinner on the table in less than 30-45 minutes (check out all the noodle dishes). I hope you try the recipes and go back again and again to make them your own, adding your own flavors and seasonings. After all, food is not new or original. It's traditional, handed down from one generation to the next from kitchen to kitchen and family to family and friend to friend. It's tried and true. Food is not just for "professionals" or chefs. It's for all of us!

Think of the best cook you know. Who did you picture? Your mom? Grandmother? Aunt? Friend? Almost universally it's not someone who makes his or her living in the kitchen of a restaurant. It's someone you love, someone you grew up with or hang out with now. It's someone you have relationship with. Someone you trust. Someone who knows you. So don't be timid. Step up. Get involved.

Think about apple pie for instance. That's nothing new but maybe having someone walk you through step by step is a new concept to you. So try it! Cooking really is easy and no ingredient in this book should be a mystery. They're all natural and normal foods grown around you and available to you; in fact, as I said before, you probably have at least 90% of them in your kitchen now. The few you don't have stocked will all be available at your local grocer. So be brave! You can do this! You just have to slow down, take time and enjoy yourself in the process of loving your people through food.

And I hope as you walk through the recipes you'll see the love it produces as you serve the people in your life. I pray you'll get rid of the hustle, if even for an hour or so. And as you lose the hustle and slow down you'll enjoy yourself and your friends and family. That you'll feel rejuvenated - physically and mentally. Regenerated. Reconciled. Those feelings will bring about health and restoration to you and in your relationships. Those are emotions I pray over all you who pick up this book.

Enjoy every taste!

Aliceson

Barry's Story

Hey Y'all!

Thank you so much for joining us on this exciting new step of our journey! My name is Barry Bales and it's my honor to tell you our story.

In 1882, my great-great grandfather, William Clark Dixon Hutton, purchased this land. Upon the marriage of his daughter, Ellen, to Porter Monroe Boles, he turned the farm over to them. Porter and Ellen raised burley tobacco, livestock, corn, wheat and 12 kids. One of them was my grandfather, Fred Marshall Boles. Known throughout the community as "Red", he was my hero. As far back as I can remember, he and my dear grandmother Mary May would pick me up on the day school dismissed for the summer. I would stay with them here on the farm until the day before school started back in the fall. Fred and Mary May's only child, Patsy (my mom), moved away from the farm when she got grown. She married my dad, W.T. Bales, and went to work for Eastman Kodak in Kingsport, which is where I lived. But it's not where I grew up...

The farm was where I wanted to be. I followed Red everywhere he went. As a little boy, I rode on his knee on the old 8N Ford tractor. We rode horses together. We checked on the cattle together. We made garden together. We sat on the bench outside Glenwood Store, whittling and lying with all the other loafers. Red was also a master carpenter and I went with him on the job on a regular basis. I spent every minute I could on the farm up until I was in my teens.

Now, we all know what they say about change. The only constant in this ol' world is change. I went off to college. And Red and Mary May got older. And I got a career as a professional musician. And Red and Mary May got older. And I traveled for a living, so my visits to the farm were fewer and further between. And in 1992, Red passed away.

Mary May stayed on at the farm until a stroke forced her to move away. I was living in Nashville, touring 250 - 300 days a year as a member of Alison Krauss & Union Station. But it just never felt like home. I missed the mountains. I missed the farm. So I moved back, into the old home place that was built in 1888. Where Porter and Ellen raised 12 kids. For the first handful of years I continued to tour heavily, so farming was never an option. I managed the place as a hunting property. And then one day, our old friend Change came calling once again.

On a blind date (the first and only for either of us), I met and would eventually marry the super-hot, intelligent, encouraging, vivacious author of this cookbook. At the time, she was a Physical Therapist and she had no idea what she was getting herself into. Neither did I, for that matter. Fast-forward a few years to the birth of our son Marshall. Yes, he's named after Fred Marshall Boles. I started to realize that I wanted him to have at least some of the experiences I had as a boy. I wanted him to be raised around animals, to know where his food comes from, and to have a strong work ethic. So we bought a few chickens. Then we bought some goats. Then we got rid of goats. And we bought some more chickens. And some pigs. And some cows. What started out as an effort to raise our own food, snowballed into a fledgling farm business that grew a little more each year.

Which brings us to where we are now. My touring schedule is infinitely easier and more manageable, so I am home A LOT more than in years past. We are raising grass-fed/grass-finished beef, pasture-raised pork, pasture-raised chicken and free-range eggs, using regenerative agricultural methods that build healthy soils, grow better forage, and produce happy animals. We supply retail customers, restaurants, and small grocery stores throughout East Tennessee. Now we find ourselves working harder than ever on the farm to keep up with demand - but we are thankful! God has so richly blessed us here that it's hard to believe. This cook book is an extension of this chapter of our lives.

When I consider what Red might think about all of this, I'm sure his first reaction would be to tell me I need to quit farming and concentrate on an "easy" job. "Quit working yourself to death, boy." I can hear it now. But I'm also pretty sure there would be a part of him deep-down that would be very proud of that fact that I'm here, trying my best to follow in his footsteps.

Fred Marshall "Red" Boles

Barry and Red on the Ford 8N tractor

Red, Mary May and Patsy Boles

A threshing crew in the back field

A note about the foods used in this book:

Obviously we care a great deal about the foods you eat and are feeding your loved ones. And obviously it matters a great deal to us where your food comes from, how it arrives to your home and even the life it lived before becoming food at all. It's our life's passion to provide people with the very best food available - meaning the most nutritious and delicious food available anywhere. Period.

Our lives now are dedicated to the animals we raise and the products produced here on our farm. This is not some hobby, some fly by the night momentary curiosity. Providing the best possible lives for animals is our life. We believe strongly that if we (and in extension you) are going to consume meat and choose to be omnivores we have a responsibility to provide whatever we're eating with the life God intended for that animal to live. We raise all our animals in fields. They are given the life they were intended to live. That means cows eat grass and only grass. Chickens pick around for bugs and eat clover as well as a supplemental feed that's designed specifically for their little bodies and their simple digestive system. And pigs root for radishes, turnips and pumpkins while foraging on grass and hickory nuts, acorns and walnuts all the while also receiving a nutritious and natural supplemental feed to give them the calories and nutrients their growing bodies need. No animal on our farm receives antibiotics, steriods or growth hormones. If given the life God intended animals don't need such things. They get those antibiotics and hormones in commercial feed lots because they are forced to live an unhealthy life and eat unnatural foods. Because of their lifestyle they are forced to take these supplements in Concentrated Animal Feeding Operations (CAFOs). But it doesn't have to be this way!!

All this is to say it MATTERS where your food comes from! It matters the life it lived and the manner in which it was treated. It matters how it ate and what it ate. It matters from animal to farmer to consumer. It matters. And the great thing is you are the boss of it all! You get to choose! You are the boss! So get choosy! Be picky!

We understand not everyone reading this book will purchase meat or eggs from our farm. While that is what we eat it's just not feasible for everyone. If you can't get our products please support someone local to you! There are tons of farmers all across this beautiful land. They sell directly to customers and also at farmers' markets. Please visit one and talk to them. Find out who you trust and support them year round. It makes such a difference. To everyone.

And one last note on this subject. If you use whole foods raised in a natural way and given the chance to live the life the animal was meant by God to live you will see such a difference. You'll have flavor you never knew was there. Your meat won't need a ton of "flavor packets" or spice envelopes you can pick up in the grocery store. You'll find out that chicken really does have a taste! Not like "stuffing from a teddy bear", which is what Julia Child likened chicken to when she moved back to the US after living in France for years. She was horrified at the commercial chicken industry and sought out whole chickens raised on pasture for the high quality taste and meat. Like Julia you'll find you can showcase the meat and let the quality speak for itself.

So with all that said we recommend some foods and ingredients in this book. Unless otherwise stated, we use:

Whole chicken raised on pasture. Obviously I agree with Julia Child in regards to the teddy bear stuffing. Before we raised our own chickens I really, really didn't think there was a difference and I really, really was afraid to even try it. I had in my mind tough, old chicken that would be just terrible with little meat. We grew different breeds and researched and tested and tasted. We have experimented with time tables, rotations, feed and seasons of the year and believe we've come upon the winning combination. So early on our poor customers had chicken that we have tweaked and improved upon now. We now raise the cornish cross breed of chicken, which is a double breasted chicken and more like what you're used to when you buy chicken in the grocery store or a restaurant. But because we raise them on pasture with a natural feed and without growth hormones, steroids or antibiotics they grow slower and have a natural life. That gives them a better texture and taste and gives you a more nutritional product. And I believe you'll agree there isn't any comparison with the commercial poultry industry that pushes these chickens to market weight in 35-42 days in an indoor environment. For instance, research shows that pasture-raised chicken has 21% less fat overall, 30% less saturated fat and 10% less calories than conventionally-raised chicken.

Grass fed and finished beef. Whew. There are so many words and phrases that can confuse folks. Some producers try to hide exactly what they do and how they do it. That's why I always encourage people to get to know their farmers. Good farmers are authentic and transparent. They'll allow you to come and visit their farm and see their animals because they're proud of what they do and they love their livestock. You'll want grass fed and grass finished beef and even lamb. That means the animal ate grass. No corn. You don't want your cattle eating grain. That's not how God designed them and it makes them sick. That's the bottom line. It makes the cattle sick to eat grains day after day after day. That's why they are given antibiotics and steroids. And the bonus to giving antibiotics (to those producers, not us) is that antibiotics make cows grow big fast. Bigger and faster than normal. That's also a reason why cheaply produced beef is so cheap to buy. Grass fed and finished cattle take on average 24 months to come to weight. Versus 6-9 months for grain fed cattle. See the difference? Cheap and fast is not the life God wants anyone to eat. Not them. Not us.

Pasture Raised Pork. Same on the pork. Know where it's from, how it was raised and what life it lived. We're big on the pig around here. Pigs are our favorite animals to raise and we love on those piggies!! Barry painted a huge painting of Wilbur, one of our sweet pigs, and it hangs over our mantle. We also have some amazing photographs of us in formal attire with our pigs (thanks to our amazingly-talented and courageous friend and photographer, Tina Wilson). We just flat out love pigs. Pork is not unhealthy by nature. If you seek out pasture raised pork it's lean, high in protein and omega-3 and low in saturated fat and inflammatory-causing micronutrients. Pasture-raised pork is also one of the best sources for vitamins D and E as well as selenium. So don't be afraid! Get lean pork and eat well.

Meat in general. In general we use whole cuts of meat and/or whole animals (chickens). Our pork chops are bone in. I do occasionally cut chickens into pieces and they have bones attached. Our beef roasts and pork shoulders have bones. Bones are NORMAL people. NORMAL. I have friends and customers who HATE and are afraid of bones. We have bones; they have bones; all God's creatures have bones (for the most part in Kingdom Animalia anyway). So don't be afraid of buying meat with bones. It keeps them from drying out and you can save the bones for amazing uses (bone broth or even treats for your dogs).

Eggs. Marshall has been in the egg business for years now and has a waiting list for his eggs. He has customers far and wide who are kind enough to support a kid with a dream and a job. So thank you to all who support him! (And here's my plug for supporting kids in business - just do it. You won't regret helping them grow and learn and dream.) Anyway, so we get all the cast-off eggs, which is fine with us! A little crack or dirt never hurt anyone. Pasture-raised eggs have 10% less fat, 40% more vitamin A and 400% more omega-3s than conventional (ie, caged) eggs.

Butter. I use unsalted butter. Martha Stewart taught me that when I bought my very first cookbook in 1993 from Sam's Club. It's better when you're baking sweets and when you're preparing savory dishes it works better to add salt when tasting at the end of the dish's preparation anyway. Salted butter does have a slightly longer shelf life but hopefully by this point in time we've all learned not to fear the butter but fear the margarine. Dairy that doesn't need refrigeration???? That is crazy.

Oils, vinegars and seasonings. I use extra virgin olive oil when I can get it and olive oil when I can't. But I do get high end olive oil and vinegars. I buy them from a store about an hour away from the farm, in Gatlinburg, TN. The store is called Zi Olive and in the Village Shopping Center in the middle of town. Zi Olive is tucked away in the corner and owned and managed by the sweetest, most gentle and generous man, Don Goins. And Don is so knowledgeable! He has the heart of a teacher and is the best marketer I've ever known. I highly HIGHLY recommend his oils and vinegars. They're the best around, in my opinion. If you can't run over to Gatlinburg (which I recommend, especially in the fall or spring - it's BEAUTIFUL), find some good quality oils and vinegars near you.

I tend to use fresh herbs anytime I can but when they're not growing in my garden I use dried. The dried herbs will last about 6 months.

Kitchen and Larder Essentials

Here are my recommendations for the basic items you need to get through this cookbook and any tasks in the kitchen.

Cast Iron Skillets: Woah Nelly! This is my top recommendation. If you don't have a cast iron skillet please get you one! I cook in my cast iron skillets every time I make a meal I think. I have a lot of cast iron cookware but if you are just getting started I recommend at the very least a large skillet (10 inch) and a small skillet (8 inch). I have pieces from both my grandmother and Barry's grandmother and then several others that I've picked up over time. I use my skillets for cornbread, biscuits, Dutch babies, nachos, cobblers, pies, roasting chicken, roasting veggies, toasting nuts and every other things you can think of. They are easy to use, easy to clean and just so homey. Unlike most cookware cast iron holds up and actually gets better with time.

Other cookware: You know there's a lot of debate about cookware. I have my grandmother's set and it's fine. It's not expensive and it gets used a lot. It gets the job done. I also have some Lodge Enamel-coated cast iron pieces that I use the most. They're colorful and fun, plus they weigh enough that you don't need to lift weights after you cook. You've done weight training for the day! Basically I think every kitchen needs a large and small sauce pan, a sauté pan and Dutch oven. Those are my go-tos. I also have a small sauté pan that I fix a veggie omelet in for lunch everyday. It cooks 2-3 eggs perfectly. I do recommend getting the pans that can go from stove top to oven. Some of my pieces can't do that and I really wish they could. It's an efficient way to prepare meals.

Poultry Shears: I know you think you won't use them but you will, trust me. So just grab a pair next time you're at Williams-Sonoma. You'll be glad you did. You can use them for so many tasks in the kitchen.

Food Processor: Barry gifted me a food processor one year for Christmas. I didn't use it for a long time but then I flipped the switch and now I couldn't live without it. I use my processor multiple times a week. The southwest spinach dip comes together in mere seconds with a processor.

Juicer: I don't have a lot of kitchen gadgets. I don't have an air fryer or instant pot. I have a crock pot but don't use it much. But I have splurged on juicers. We've gone through several in our marriage actually. We don't juice every day but do try to juice several times each week. My go-to juice is a beet/carrot/apple juice or celery/cucumber/apple. Research shows people who average 3 servings of fruits and veggies each day look 11 years younger than they actually are. I don't know if it works for me but I'll try it! And people sometimes poo poo juicing because "it's not as healthy as eating the entire veggie" but listen I don't eat beets. So if I'll drink a beet juice that's better than a Big Mac.

A good blender: Did you notice the "g" word in there? I've had good blenders and cheap blenders. The cheap ones are inexpensive for a reason - they're cheap and don't work. They especially won't last. Get the good one.

Ice Cube Trays: Because trust me your fridge that makes ice is gonna crash on Thanksgiving. Or you'll run out in the middle of a party while you're making cocktails. They're also great for freezing pesto or herbs for the winter. Just cut dried herbs and cover with olive oil and freeze. They'll be great in soups and pasta dishes all winter long.

White Rice Flour: keeps pizza crust and bread loaves from sticking to the pans and won't change the flavor or consistency like cornmeal does.

King Arthur Flour: that's my go-to bread flour

Mason Jars: if you stick with either wide mouth or regular jars you won't ever have to wonder what size lid you need. I have both but if I was smart I'd only do wide mouth. I think they're easier to pour into and stack.

Wooden Spoons: I think they look good, feel good when I'm using them and they do the job when a toddler is acting out. Scientifically speaking they're non-reactive so won't alter acidic foods like metal spoons can.

A good mixer: it'll make baking a breeze!

Appetizers

I was hesitant to put this in the cookbook. Not because I don't love it or think it's worthy (it is!) but just because it's not really a recipe. It's more of an idea.

We obviously have a lot of pork on the farm and we love it immensely. In my educated opinion Barry's pork barbecue is some of the very best around. He does an amazing job at smoking a pork shoulder! Truly. It is the best around here. So we're always experimenting with different ways to use the barbecue, especially when we just have a little bit left. This is a great way to enjoy the little bits you have remaining.

Pork Barbecue Nachos

serves 2-4

hands on time 10 minutes

cooking time 10 minutes

total time 20 minutes

Enough tortilla or potato chips to cover a 9x13 inch pan (I do homemade potato chips here. They're easy. I promise. Here's how you do it: slice potatoes very thin - food processor works well. Dry them really really well and deep fry them on 375 degrees for 3-4 minutes per batch. Salt while hot and put them on your pan).

Leftover pork barbecue

1 fresh jalapeno, sliced

1 cup Monterrey Jack cheese, grated

barbecue sauce

1 tablespoon fresh cilantro leaves, shredded

Midland White Pepper sauce

Once your pan is covered with a thin layer of chips, sprinkle barbecue, jalapeño and Monterrey Jack cheese on top. Place in an oven with the broiler on and broil until the cheese is melted and bubbly. Remove from oven and drizzle barbecue sauce and white pepper sauce on top and sprinkle cilantro cheese on the top. Eat outta the pan!

I used to buy frozen pizza rolls in college and our early married life. Now they just taste like the bottom of a tennis shoe. Am I right? Barry said the other night he has no idea what they've done to them but it's not good (and yes, sometimes I do still serve them to the boys - don't judge).

Homemade pizza rolls are to frozen what Vera Wang is to K-Mart or a Porsche to an Escort. Or something great compared with something not so great. You get the picture. Yes they are intense to make. Yes they take time. And no I don't do them a lot. Usually just when my nephew is on the farm or for a big game. It's not an everyday thing or even every week thing. These pizza rolls are involved, somewhat difficult, messy and in the end worth *EVERY SINGLE MOMENT* of your time. Trust me. Make these for yourself and your loved ones. You'll never regret one second of it. I promise.

The Most Amazing, Awesome, Life Changing Pizza Rolls

serves 4-6

hands on time 30 minutes

cooking time 5 minutes per batch

total time 35 minutes

1 package egg roll wrappers
1 cup mozzarella cheese, shredded
1/4 pound Bales Farms sweet or hot Italian sausage, cooked
water for your fingers or brush
vegetable oil for frying

Marinara sauce:

1 (28 oz) can crushed Roma tomatoes
2 cloves garlic, minced
handful fresh basil leaves, cut into strips
salt to taste
1/2 teaspoon crushed red pepper (optional)

Heat the oil in a large skillet to medium high heat.

Take the egg roll wrappers out of the package and cut them in the middle, making 2 rectangles.

To assemble the pizza rolls put a very small amount of marinara sauce, sausage and cheese in the middle of each roll. Brush or smooth with your finger water along the edges of the wrapper and fold to make a small rectangle. Make sure you seal the pizza roll completely. You may want to use a fork to help with this. Continue with all the pizza rolls.

Once the oil is hot enough (if you have a thermometer it should read 350-375 degrees. I never have a thermometer so I flick some flour in and see if it bubbles. If it does the oil is ready.) Fry the pizza rolls in small batches until golden brown.

Serve while hot with additional marinara sauce.

Note - *DO NOT OVERFILL the pizza rolls!!!!! It takes a tiny amount to fill each roll.*

If there's one recipe I'm known for this is it. My white cheddar pimento cheese. It is ALWAYS in our fridge and usually is gifted to someone every week. It's great for new neighbors, new mommas, date night with wine, tailgating with beer, on grilled cheese sandwiches, on a charcuterie tray. You get the drift. My pimento cheese has been a staple for me but it is different than other recipes as its not a spread. There is no filler in this recipe. It's all real ingredients and is just about perfect.

White Cheddar Pimento Cheese

serves 8-12

hands on time 15 minutes

total time 15 minutes

2 (6oz) blocks white cheddar cheese, grated (DO NOT BUY THE BAGGED CHEESE)

1 (4oz) jar pimentos

1/2 to 3/4 cup mayonnaise

8-10 shakes Midland pepper sauce (or other hot sauce)

salt to taste

Mix pimentos with juice, mayo, hot sauce and salt. Add in cheddar cheese. Stir to combine. Store in the fridge for 2 weeks but it probably won't last that long.

Note- I roast a jalapeno pepper and dice it when Barry is eating the whole batch. It's often too spicy for my mom and friends' kids but he loves it.

Woah. These babies pack a punch. They are perfect for charcuterie or a picnic or even on a sandwich. I will warn you, though, they have a very strong smell when you open the jar - that's just radishes. They taste great and are totally worth the odor!

Pickled Radishes

serves 8-10 at least

hands on time 5-10 minutes

cooking time 5 minutes

total time 15 minutes

1 bunch radishes (1/2 pound), sliced thin

1/2 cup apple cider vinegar

1/4 cup water

1/2 cup sugar

1 teaspoon salt

1 teaspoon mustard seeds

1 teaspoon black pepper

Place radishes in a mason jar. Bring water, vinegar, sugar, mustard seeds and pepper to a boil in a small sauce-pan, stirring constantly. Pour over the radishes and let cool, swirling to cover the radishes. Stir in a fridge for 1 month.

This is another great addition to a charcuterie board. It's also fabulous on street tacos, nachos, barbecue sandwiches or any other concoction. It's super easy and beautiful. The onions retain their beautiful pink color and it just pops on every dish or tray. It's so easy and lasts for at least a month in the fridge. I'd recommend you make them today!

Pickled Red Onions

serves 8-12

hands on time 10 minutes

cooking time 5 minutes

total time 15 minutes

1 red onion, sliced thin

3/4 cup good quality red wine vinegar

3/4 teaspoon Kosher salt

Slice the onions and place into a small Mason jar. Bring the vinegar and salt to a boil, stirring constantly. Pour over the onions and swirl to cover. Refrigerate for a month.

Note- I used to keep these onions sliced but now I also cut them into cracker-sized pieces to allow for more civilized snacking.

This sweet little appetizer is a favorite of our friend, Barry Scott. Barry and Barry grew up together and have been friends from elementary school. They've seen a lot of the world together, even winding up in the same hotel in New York City one night when Barry Scott was traveling for work and Barry Bales was on tour. Barry and Jennifer are some of our dearest friends and it's a privilege to know our sons are good friends, too. It's always sweet to have them around and every time they come I always serve this dip for the Barrys.

Southwest Spinach Dip

serves 4-8 or 2 Barrys

hands on time 10 minutes

cooking time 0

total time 10 minutes

1 (8 oz) block cream cheese, softened

1/2 cup mayonnaise

1 handful fresh cilantro

1 tablespoon fresh lime juice

1 fresh jalapeno, cap removed

salt and cumin to taste

8 oz fresh spinach

In a food processor combine cream cheese, mayo, cilantro, lime juice, salt and cumin and jalapeño. Process until smooth (will be a very light green). Add spinach and pulse until blended (should be a vibrant green color). Taste and adjust as needed (you'll need more salt than you think probably!). Serve with tortilla chips.

I can everything. If it will keep still long enough I'll can it. Except salsa. I haven't had a canned salsa I like because I can't stand the taste of vinegar in my salsa (although I'm sure if I had yours it would be amazing and you would change my mind). So I don't can salsa. I do this version instead. I originally had something like this at my friend, Tina's house, although she puts corn in hers I believe. So I've tweaked it and tweaked it over the years and here's my go to recipe. I hope you like it!

Salsa

serves 8-12

hands on time less than 5 minutes

cooking time 0

total time less than 5 minutes

2 cans Rotel (Chipotle is the best)

1/2 cup onion

1 jalapeño, cap removed

handful fresh cilantro

cumin to taste

pinch sugar

1 teaspoon salt

1 teaspoon lime juice

Add everything to a blender and let it rip! There's no need to chop anything in this recipe and it's a good place to even use the stems of the cilantro. Taste before you pour it out of the blender because you may need more salt, sugar or cumin. Serve with tortilla chips.

I debated where to put these tacos. We have them usually on Friday nights with chips and cheese dip, salsa and guacamole. The tacos always take on a different flavor because I use whatever meat I have left over from the week. It's a revolving door of chicken (from the buttermilk chicken recipe), chorizo, steak or ground beef. They're super simple, come together in no time flat, use leftover ingredients and allow you to enjoy a La Paloma on the porch with your people! You just can't beat that on a Friday night. Everybody deserves an easy Friday evening.

Also this "recipe" is more of a to-do list, since you should have most ingredients already made or just need to chop them up.

Street Tacos

serves 4-6

hands on time 10-15 minutes

cooking time 5 minutes

total time 20 minutes

White corn tortillas

Choice of meat (chicken, chorizo, steak or ground beef, cooked and left over from another meal throughout the week)

spinach or lettuce, shredded

Monterrey Jack cheese, grated

fresh jalapenos, sliced

fresh tomatoes, diced

red onion, diced

lime juice

pico de gallo (either made from the above ingredients or bought)

Place oven on broil and add tortillas to brown. Once browned, place in a taco holder to form as the tortillas cool. (This is a great way to save on calories and grease. Tortillas will keep the shape you place them in as they cool. If you don't have a taco holder I recommend you get one! My dad, who makes our charcuterie boards, offers taco holders if you're in need.)

Once the tortillas/taco shells are formed, add the meat, shredded greens, tomatoes, jalapeño peppers, onions, lime juice and cheese. Drizzle with taco sauce and Midland Ghost White Pepper sauce as desired.

Serve immediately as everyone else in your house is already chowing down on the chips and dips.

This is another old recipe. Vintage. But still good. This was my first "big girl" attempt at cooking and entertaining. It's still good and I still pull it out every now and again. It's so so good and everybody likes it. Even kids who would shun all the ingredients alone love them together. And even though it might feel like you're on a first date at TGI Friday's in the 1990s and you suddenly start craving a Cosmopolitan, it always pleases.

Spinach-Artichoke Dip

serves 10-12

hands on time 15 minutes

cooking time 20 minutes

total time 35 minutes

2 cups fresh spinach leaves, torn

14 oz artichoke hearts, drained

1 (5.5 oz) container soft spreadable cheese, garlic and herb flavor

1 cup Mozzarella cheese, grated

8 oz sour cream

1/2 cup mayonnaise

2 oz pimentos, drained

3 slices bacon, cooked and crumbled

Preheat oven to 400 degrees.

Stir all ingredients in a large bowl and spoon into an 11x7 baking dish. Bake for 20 minutes or until bubbly and golden. Serve with tortilla chips or, if you're feeling really saucy and formal, baked baguette slices.

I LOVE rosemary and lemon together. Really rosemary and any citrus. It's such an amazing combination. I will warn you to go light on the rosemary, though. If you're too heavy-handed with rosemary (like lavender) you'll end up with soap.

Here's an amazing summer concoction that will be a hit at any gathering.

Cherry and Rosemary Lemonade

serves 8-12

hands on time 5 minutes

cooking time 5 minutes

total time 10 minutes

1 cup sugar

1 cup water

1 sprig fresh rosemary

1 quart homemade lemonade (or store bought if you're in a pinch)

handful fresh cherries

1 lemon, sliced

In a small saucepan over medium heat combine sugar and water and stir until sugar is melted. Add rosemary and let cool. Remove rosemary.

Add 1/2 cup simple syrup to quart of lemonade and pour into a large jar. Garnish with cherries and lemon slices.

Note - this pairs nicely with a splash of champagne for a summer evening dinner party or picnic.

Soups

I love this chowder! It's one of my favorites for a crisp Fall afternoon. It's just so lovely and colorful and savory and cheesy and comfortable. There's just something about this yummy melding together of flavors. I really hope you try it and love it as much as we do. It's one of my top requested recipes ever.

Cheesy Chicken and Corn Chowder

serves 6-8

hands on time 15 minutes

cooking time 30 minutes

total time 45 minutes

3 slices bacon, cooked and crumbled

1 red onion, diced

1/2 green bell pepper, diced

2 cloves garlic, minced

2 medium potatoes, diced

3 cups bone broth

1 1/2 cup frozen corn

3 cups whole or 2% milk

3 tablespoons all purpose flour

2 cups dark and white meat from a leftover buttermilk chicken

1/2 cup colby cheese, grated

salt and pepper to taste

In a large Dutch oven or soup pot cook bacon and remove, saving the grease. To the grease add onion, green bell pepper and garlic and sauté for 5 minutes, or until tender. Next add potatoes and continue to sauté for 5 minutes. Once the potatoes are beginning to brown add bone broth and bring to a boil, then reduce heat and simmer for 10-15 minutes, or until the potatoes are tender. Stir in corn.

In a small bowl mix together milk and flour. Stir into soup and continue to stir until thickened. Add chicken, bacon and cheese. Season with salt and pepper.

Serve with a sprinkle of cheese on each serving.

See? I've got a lot of recipes using leftover chicken and bone broth. They're all nutritious and simple, easy to make and come together quickly. I've been making this one since Barry and I got married. It started as a complete mess up. I was trying to make a white chili but didn't have beans so forgot to add those. Then I deleted some things and added some others. Over the years I've made it probably 1,000 times so it's come to be it's own thing now that we refer to as "Southwest Chicken soup".

Southwest Chicken Soup

serves 6-8

hands on time 15 minutes

cooking time 20 minutes

total time 35 minutes

1 tablespoon olive oil

1 red onion, diced

1/2 green pepper, diced

1 jalapeno, diced (remove seeds for a milder taste)

1 clove garlic, minced

2-3 tablespoons chili powder

1/2 teaspoon cumin

4 cups cooked buttermilk chicken, chopped

4 cups bone broth

1 (12 oz) bottle chili sauce

1 (15 oz) can fire roasted diced tomatoes

salt and pepper to taste

Garnish:

fresh cilantro and avocado

In a large Dutch oven, heat oil on low to medium heat. Add onion, green bell pepper, jalapeño and garlic. Sauté for 5 minutes or until tender and fragrant. Add chili powder and cumin; stir to coat veggies. Add chicken and stir to combine. Add broth, chili sauce and diced tomatoes and heat thoroughly. Add salt and pepper to taste and serve with a garnish of fresh cilantro.

I was introduced to this soup in the Miami airport of all places. I know. Crazy. We were traveling and Barry was feeling a little under the weather so got a spicy soup to help him get over the hump and get back home. He let me taste it and I couldn't believe how great it was! I knew I had to learn how to make that soup and started writing down all the tastes I could think of while we waited on the plane. Then I googled it and played around with the recipe until I came up with one we all really like. It's not exactly what we had in Miami but we sure do love it. I've shared the recipe with friends who've also enjoyed it so we're not as crazy as it seems.

Thai Chicken Soup

hands on time 15 minutes

cooking time 25 minutes

total time 40 minutes

2 tablespoons butter

1/2 red onion, chopped

1/2 red bell pepper, chopped

2 small carrots, sliced

5 mushrooms, sliced

1 cup uncooked brown rice (you can use white rice for sure! But brown rice tastes the same here and is a bit healthier)

2 tablespoons Thai red curry paste

2 cups cooked Bales Farms chicken, chopped

5 cups bone broth

2 teaspoons (or more, depending on how spicy you like your soup) garlic chili sauce

2 cups heavy whipping cream

fresh cilantro

Melt butter in a large soup pot and add onion, pepper and carrots. Sauté for 5 minutes. Add the mushrooms and rice and continue to sauté for 2 minutes. Add the curry paste and stir to combine. Pour in the bone broth and bring to a boil, then reduce to simmer until the rice is tender. Once the rice is tender add the chili garlic sauce, chicken and whipping cream. Stir until all flavors are combined. Ladle into serving bowls and garnish with chopped fresh cilantro.

By sautéing the rice and adding spice before liquid you'll get a bolder flavor in the rice.

My dad had this on vacation one time and I didn't even eat it but was mesmerized by the appearance and aroma. I knew I had to make it when I got home. So I've played around with it for years now and realized it was a combination of canned corn and fresh (or frozen in the winter) corn. It's quick, simple, feeds a crowd and keeps everybody happy. It's especially great on a day you want to just throw things in and let them go, it's literally that simple. I fix this so we'll have a quick and filling lunch on weeks when I know we'll be super busy.

Double Corn and Bacon Chowder

serves 6-8

hands on time 10 minutes

cooking time 30 minutes

total time 40 minutes

4 slices bacon, cooked and crumbled

1/2 red onion, diced

2 medium potatoes, diced

1 cup bone broth

2 cups whole or 2% milk

1 (15 oz) can creamed style corn

1 cup fresh or frozen corn

salt and pepper to taste

Cook the bacon in a Dutch oven until crisp and remove, saving grease. Add onion and saute for 5 minutes or until tender. Add potatoes and sauté for 5 minutes. Add bone broth and bring to a boil then reduce to a simmer and cook until tender, about 10 minutes. Add milk and corn, stirring to combine. When heated through add crumbled bacon back to Dutch oven and season with salt and pepper to desired taste.

So. Our chili. It's debatable. I'm not a big bean person. Scratch that. I'm not a bean person. And as Barry gets older he's not either. So if I make this chili for us (and I do about every other week in the fall and winter) I don't put beans in it. I'm not a monster, though, so I do add beans if I'm serving it to company, but I've found a lot of people secretly don't like beans in their chili either. If you're like me, rejoice! Here's a recipe you'll love and can embrace any day of the year.

This is great with a big plate of nachos or a loaf of cheese bread. . . .

Bales Farms Chili (with or without beans)

serves 6-8

hands on time 15 minutes

cooking time 30 minutes

total time 45-60 minutes

1 pound Bales Farms ground beef

1 red onion, diced

1/2 green bell pepper, diced

1-2 fresh jalapeño peppers diced, with seeds removed if desired (will result in a milder chili if removed)

3-4 teaspoons chili powder

1 teaspoon cumin

salt and pepper to taste

1 (15 oz) can fire roasted diced tomatoes

1 (15 oz) can tomato sauce

1 (15 oz) can kidney beans, drained and rinsed twice (optional people!)

Brown the ground beef in a large soup pot and add onion, bell pepper and jalapeño pepper as the beef is browning. Once brown, add chili powder, cumin and salt and pepper and stir to coat all the beef and veggies. Next add the tomatoes and tomato sauce and simmer for 20-30 minutes. If you're adding beans this is the time! The flavors will blend as the chili sits in the pot.

Garnish with grated cheddar or colby cheese and/or sour cream.

*Note - sometimes canned tomatoes have a little bitter turn to them. That taste is easily overcome with a pinch of sugar. I tend to add it to dishes with canned tomatoes. Also, I know this seems like a lot of seasonings so taste as you go with the chili powder and cumin. We like a lot of flavor in our chili so add 4 teaspoons.

Mains

I could eat pasta of some sort every day of my life. I love pasta. In any form. I have started making my own pasta in the last few years and it's life changing. I can't make it from scratch every time we have pasta but when I get to slow down and make it by hand it's always a welcome surprise. It's like a long exhale at the end of the day. It just makes life better somehow. Knowing someone took that much time and effort for you. I try to do that for my family as much as I can but, again, it's not every day.

This pasta comes together really quickly because you'll use veggies you have on hand, leftover chicken from a roasted buttermilk chicken and marinated artichokes, which pack a ton of flavor. So buckle up! You'll have this dinner on the table with crusty bread (you don't even need a salad there are so many vegetables in this dish!) in less than 30 minutes.

Chicken and Vegetable Pasta

serves 4

hands on time 15 minutes

cooking time 15 minutes

total time 30 minutes

3 tablespoons butter

3 tablespoons good quality olive oil

3 garlic cloves, minced

1 red onion, diced

1/2 cup sun dried tomatoes, sliced

1/2 cup fresh mushrooms, sliced

1 cup marinated artichoke hearts, sliced

1/4 cup black olives (optional)

1 cup spinach leaves, sliced

3 cups Bales Farms chicken, cooked and diced from a leftover buttermilk chicken

4 oz artichoke marinade from jar

1/8 cup fresh basil leaves, julienned

1/2 tablespoon fresh rosemary, leaves off stem

1 tablespoon fresh oregano leaves

1/4 teaspoon crushed red pepper

1/2 teaspoon salt

8-10 ounces pasta of choice, cooked per directions and strained from water

Parmesan cheese, grated

Prepare pasta according to package. Set aside.

In a large skillet combine olive oil and butter and melt on low to medium heat. Add red onion and garlic and sauté until fragrant, approximately 2 minutes. Add sun dried tomatoes, mushrooms, artichokes and olives, if using. Stir to combine and continue to sauté. Add spinach and chicken and sauté until the spinach is wilted, approximately 1 minute. Add marinade and seasonings and stir. Add pasta to skillet and let the flavors mix together. Top with parmesan cheese and serve.

Barbecue is controversial. Everybody loves the barbecue they grew up on and strongly believe that's the best, and maybe only, way to eat barbecue. And boy do some people want to keep their tips secret! Secret sauces, secret rubs, secret wood variations. Luckily for you we aren't like that at all!

Barry and I both grew up on Ridgewood Barbecue. Ridgewood is a hole in the wall restaurant that's been written up in Southern Living many, many times and named as one of the South's best barbecue joints. It's great and when we were growing up it was a privilege to go and get in the door. (They didn't have set hours and when they were done, they were done. Ridgewood is famous for shutting the door in people's faces. We went one time around Christmas and had 2 infants with us. It was snowing. They made us wait outside. In the snow. With babies.)

Barry and I would never do that! If it's snowing I'll let you at least wait on the porch. (Just kidding! I'd totally let you come in.) And while Ridgewood won't share their recipe we will share ours. I believe Barry has surpassed Ridgewood. It's taken about 20 years and lots and lots of trial and error but he smokes amazing barbecue and our sauce is fabulous, if I do say so myself.

Now, my sauce is a tomato based sauce, not mustard or vinegar (sorry to those folks who prefer those sauces). It's simple and straight forward and it takes an afternoon simmering on the stove but it's worth it. Truly, just dump and stir. Then keep on the lowest setting for 3-5 hours and enjoy it's thick, velvety texture on the smoked pork shoulder. Or on sandwiches. Or as a base for salad dressing. or on barbecue nachos. Or on barbecue chicken pizza. Seriously, we eat it on everything and I make it a lot. All the ingredients are handy and you probably have them in your pantry. And, yes, buying bottled sauce is easier and quicker but have you seen the list of ingredients on bottled sauce lately? Also this sauce comes together quickly and can be hotter (add hot sauce to your liking) or with more mustard (you can decreased the ketchup to 1 1/2 cups and increase the mustard to 3/4 cup). Or add more apple cider vinegar and you'll have a thinner sauce with more of an acidic punch. I usually don't like vinegar sauces but I will say apple cider vinegar goes well with pork smoked over apple wood chips. Play around with it! You'll have time to twist the ingredients as you cook. The only warning I would give is to add salt if needed at the end. Ketchup and mustard have salt already in their ingredients and as the sauce cooks the flavors intensify. So be careful with the salt, but other than that, go for it! If your family likes a different flavored sauce, play around! This is a good, basic sauce that you can build on for your taste buds.

Smoked Pork Shoulder with Homemade Barbecue Sauce

serves 10-12 at least

hands on time 2 hours

cooking time 3-5 hours

total time 6 hours

Pork shoulder rub:

3/4 cup paprika

1/2 cup sugar

1 tablespoon onion powder

1 teaspoon garlic powder

3-4 pound Bales Farms pork shoulder, defrosted

Barbecue sauce:

2 cups ketchup

1 cup water

1/2 cup brown sugar

1/8 cup honey

1/2 cup apple cider vinegar

3 tablespoons Worcestershire sauce

2 tablespoons yellow mustard

1 tablespoon onion powder

1 tablespoon pepper

1 tablespoon paprika

Mix seasonings in a small bowl. Rub the defrosted pork shoulder with the seasonings. Wrap the shoulder tightly and refrigerate overnight.

Prepare your smoker with apple wood and hickory wood chips, soaked in water. If using a Big Green Egg, use the convEGGtor for indirect cooking. Plan on roughly 2 hours cooking time per pound of meat. Once the smoker reaches 225 degrees put your pork shoulder on the smoker. I keep an aluminum pan of apple juice under the shoulder to add moisture. Smoke until the internal temperature reaches 165 degrees. Be sure your thermometer is in the thickest part of the meat near the center, not touching any bones. Wrap the shoulder in 2 layers of aluminum foil, then return to the Egg/smoker and continue to cook until internal temp reaches 195 - 200 degrees. It should fall off the bone at that time and will be obvious when it's ready. Remove from Egg/grill, wrap in a towel and place in a cooler to rest for at least 30 minutes. Remove and pull apart.

In a medium saucepan add all ingredients for the sauce and bring to a boil, then reduce and simmer for 3-5 hours, stirring occasionally. This sauce will keep in the fridge for 2 months if you can keep it around that long.

I've always said I cook like I play music - by ear. Or eye. So I use recipes as guides, but I've developed a "feel" for when the meat looks and acts right. That's all part of the fun of cooking! - Barry

We've never been huge pork chop folks. Until we had our own pork chops from our own sweet pigs. We truly, truly believe in raising animals the right way and giving them the lives God intended them to live because that's what humans are called to do. If we're going to eat meat we have the responsibility to raise the meat (or purchase it from a farmer who raises the meat) in the most natural way possible. It matters - it's obviously far better for the animal, better for the world in which we all live and more nutritious for us. We raise our pigs on pasture and in woods. We have a several areas on the farm in which they rotate, starting out in the barn as little piglets and rotating as they get bigger and prove to us they are trustworthy. Trustworthiness in a pig is huge! Little pigs grow into big pigs who are strong and stubborn. My great uncle Creed used to say his wife was "As independent as a hog on ice" and when we raised pigs I understood his statement like never before. So we have to teach them and train them when they're little or else there's no containing them when they're 200 pounds.

And then we're blessed with the best pork chops ever. Truly. The best I've ever eaten. I do recommend enjoying pork chops with the bone in the chop. This keeps the chop from drying out.

One statement I believe to be true is that if you begin with quality ingredients, including meat, you need little else. We don't use a lot of flavor additions because we just don't need it. I don't buy "flavor packets" at the store. We use high quality meat and eggs, fresh vegetables and fruit and the highest quality oils and vinegars we can find. Flavor at that point takes care of itself.

So here is how we prepare our pork chops. And by "we" I mean Barry.

Grilled Bone in Pork Chops

serves 4

cooking time 8-12 minutes

total time 20-25 minutes

4 Bales Farms bone-in pork chops, cut to 1 inch thickness olive oil

salt to taste

Rub room temperature pork chops with olive oil and sprinkle liberally with salt.

Heat grill or big green egg to 425-450 degrees. Cook on the hottest part of the grill for 1 minute and flip. Cook another 1 minute. Move chops away from hottest part of grill and cook another 3-4 minutes. Flip and repeat. Cook until 165 degrees internally. Make sure temperature probe does not touch bone.

Remove from grill, cover and rest for 10 - 15 minutes.

Serve alone or with a white barbecue sauce. We very seldom use anything on our meat other than salt, pepper, and olive because we love the flavor of it so much, and we are constantly doing "quality control" tasting. Depending on where you source your meat, it might benefit from a good brine or rub.

I have always loved a wood fire. For grilling, certainly, but definitely for smoking. I have always had a gas grill for quick "weeknight" situations, as well as a barrel smoker with side fire box for when I can take my time or for smoking. A lot of friends started getting Big Green Eggs and proceeded to try to talk me into getting one. I never had any doubt they were great, I just didn't know where one would fit in my "arsenal". Fast forward - Aliceson and Marshall got me one a few Father's Days ago. I have not used anything else since. I love it! - Barry

Perfect Grass Fed Steak

Ingredients

- Bales Farms grass fed/grass finished steaks
- Good quality salt
- Olive oil

• Thaw steaks

• Rub with olive oil and good quality salt

• Heat BGE to 650 - 700 degrees

• Cook steaks for 1 minute & 10 seconds over direct heat, then rotate 180 degrees and cook for an additional 1:10

• Flip steaks over and repeat step 4.

• When finished, put a pat of butter on each steak.

• Cap egg and close vent completely, let sit for 4 minutes.

• Remove steaks to plate and cover with foil.

• Rest steaks 10 minutes at room temperature.

• Serve and enjoy!

This will get you Medium Rare. We do not recommend cooking grass finished steaks beyond Medium.

White Steak Sauce:

1/2 cup Blue Plate or JFG mayonnaise

2 tablespoons Midland White Pepper sauce

1 garlic clove, minced

1 tablespoon Creole Mustard

1 teaspoon prepared horseradish

1/2 teaspoon parsley

salt and pepper to taste

Mix everything together in a small bowl or mason jar. Refrigerate until ready to use.

Who doesn't love a baked pasta dish? They're so easy for a weeknight meal or to take to a friend who's had a new baby or come home from the hospital. Throw this on the table on a Tuesday night with sliced homemade bread and a salad and your people will love you for at least the night.

This recipe is for my baked pasta using our sweet Italian sausage, which is one of our best sellers and a universal favorite. The sausage is perfect in pasta, on pizza and sandwiches. I use it to make meatballs and serve those at parties for young and old alike. You just can't go wrong with a simple dish, baked in an oven using good, nutritious and delicious ingredients.

Baked Penne Pasta

serves 4-6
hands on time 15 minutes
cooking time 20 minutes
total time 35 minutes

1 pound Bales Farms Sweet Italian Sausage
Tomato/marinara sauce:
olive oil
1/2 red onion, diced
2 garlic cloves, minced
2 (15 oz) cans fire roasted diced tomatoes or 1 (28 oz) can crushed tomatoes
1 tablespoon fresh rosemary, removed from stem
1 tablespoon fresh oregano leaves, removed from stems
1 tablespoon fresh basil leaves, thinly cut
crushed red pepper
8-12 oz penne pasta, uncooked
1 cup fresh spinach leaves
1/2 cup fresh mushrooms, sliced
1/2 cup red onion, diced
1 cup shredded Mozzarella cheese
Topping:
1/2 cup shredded Mozzarella cheese

Preheat oven to 400 degrees.

In a large skillet cook sweet Italian sausage until no longer pink.

In a sauce pan heat olive oil on low to medium heat and add onion and garlic. Sauté for 4-5 minutes until fragrant and tender. Add canned tomatoes and fresh herbs as you prefer (I use at least a tablespoon of all the fresh herbs and 1/4 teaspoon crushed red pepper). This is a fabulous marinara sauce that you can use anytime and anywhere. It's my marinara sauce for pizza and all my pasta. It's always in my fridge because I use it every week and just make more as I run out. It only takes about 10 minutes. Once the marinara sauce is ready add spinach leaves, onions and mushrooms and heat thoroughly.

In a large pot cook pasta according to directions. Strain and pour into a 9x13 inch baking dish (or smaller and deeper dish). Pour sausage, marinara sauce with veggies and cheese in and stir to combine. Top with cheese and bake for 20 minutes.

This vegetarian-friendly main dish is a cinch to pull together on a weeknight. It can also be made friendly to all your vegan loved ones! If you don't have all these veggies, no worries. Use what you have. I have substituted cabbage, sweet potatoes and squash/zucchini in this dish. You truly can't go wrong. Just use what you have. Your hands on time with this dish is minimal. The main time component is for baking. So sit back, let the oven do your work and get to helping with that homework!

Baked Quinoa

serves 4

hand on time 10 minutes

cooking time 40 minutes

total time 50 minutes

1 cup quinoa

2 cups water

2 potatoes, cut into 2 inch squares

good quality olive oil

1 cup fresh spinach leaves

1 Roma tomato, diced

1 cup red onion, dices

1 cup red bell pepper

2 garlic cloves, minced

1 cup mushrooms, sliced

1 cup chopped fresh broccoli

salt and pepper to taste

Preheat oven to 400 degrees.

Combine quinoa and water in a small saucepan. Bring to a boil and then reduce heat. Cover and let sit for 15 minutes or until water is absorbed.

While the quinoa is cooking, roast potatoes in a cast iron skillet with oil until the quinoa is ready. The potatoes will cook 50% of the way on their own and then continue cooking in the next step (hold on!).

Take the skillet out of the oven and add the quinoa, veggies and salt and pepper. Stir to combine and return to the oven to roast for 20 minutes, or until the potatoes are tender and the quinoa is crisp.

Top with Midland White Pepper Sauce (see recommended supplies list) and serve straight out of the skillet!

Here it is - the famous yet mysterious buttermilk chicken. I can't tell you how many calls, texts, messages and shouts I've received over the years about the mysteries of a whole chicken. Let me say this right here, in print, THERE IS NO MYSTERY ABOUT PREPARING WHOLE CHICKENS!!!

It takes the same amount of time to prepare a whole chicken as it does parts and pieces and yet the reward is so much more with a whole chicken! Truly. It's so so easy. And you have meat left over for multiple meals (see the casseroles, pastas, pizzas, salads, sandwiches and soups listing) plus the bones for a nutritious broth. Working with whole chickens is cheaper, easier and much faster over the long haul. You roast it once and eat multiple times. I promise.

Whole Roasted Buttermilk Chicken

serves 4-8

hands on time less than 5 minutes

cooking time 1 hour

total time 65-70 minutes

1 (3-4 pound) whole chicken, preferably Bales Farms chicken

whole milk or buttermilk (preferred)

Kosher salt

ziplock bag

Put the whole chicken in a ziploc bag (I also place the bag in a bowl in case there's a hole in the bag). Put a good amount of salt in the bag with the chicken and cover with milk until most of the chicken is covered but you can still seal the bag. Put in the fridge for 8-24 hours.

Remove the chicken from the bag and rinse. Place in a cast iron skillet breast side up. Sprinkle with additional kosher or preferred salt (I recommend pink Himalayan salt for this step). Roast in an oven at 410 degrees for 10 minutes and reduce heat to 350 degrees for 50 minutes or until the internal temperature of the chicken is 165 degrees.

Here are some tips to help you out:

1. Put the legs in first. The legs and thighs are the darkest meat of the chicken and therefore take the heat the best. They should be in the hottest part of your oven (which will be the back, opposite the door)

2. If you don't want to turn your oven down in the middle of roasting set it either at 350 degrees for an hour or 410 degrees for 40 minutes or so.

3. If your chicken is getting a little brown on top cover with aluminum foil.

4. Once you remove the chicken from the oven let it rest in the skillet for 10 minutes, then carve in the skillet and let the meat set in the gravy that has been made while roasting. It will give you the most tender and delicious chicken you've ever eaten.

5. If you can't get a Bales Farms chicken I urge you to find a local chicken farmer and buy your chicken from them. Raising meat chickens is hard work! It's the hardest work we do but we do it because the difference between a pasture-raised chicken eating all natural ingredients and given a happy life versus an animal in a shed eating whatever they eat (I can't even think of a word for it) is life changing. Truly. There is no comparison. So please, support your local farmer.

6. This chicken will be the most tender, juicy and delicious chicken you've eaten in a long long time, if not your whole life. I guarantee it. And it's so easy. Plus people are always impressed and mystified that someone can still do this. Although all our grandmothers are rolling over in their graves now because they all did it weekly and it was no big deal. And usually they roasted it after killing it and plucking it's feathers... Our grandmothers were amazing. If you still have yours please go hug her and say thank you for feeding you all those years.

This is a casserole I love so so so much. I could personally eat this casserole every week of my life, and I'm not that into casseroles (so if I love it you know it's good). This casserole is truly a one dish dinner - it's packed fully of veggies for you and enough cheese to make the kids forget. Plus POTATO CHIPS on top. Need I say more?

Cheesy Chicken Casserole

serves 6-8 at least

hands on time 25-30 minutes

cooking time 40-45 minutes

total time 1 hour

2 tablespoons butter, melted

1 medium red onion, chopped

1 package chicken flavored rice

3 cups cooked Bales Farms chicken, chopped

1 stalk celery, chopped or sliced

1 1/2 cup frozen peas

1 cup mayonnaise

1 (10 3/4 oz) can cream of chicken soup

1 (4 oz) jar pimentos, drained

6 oz cheddar cheese, grated

1/2 cup sliced almonds (optional)

3 cups crushed wavy lays, crushed

Preheat oven to 350 degrees. Melt butter in a skillet. Add onions and sauté until tender (about 5 minutes). Remove onions and prepare rice in the same skillet according to the package (why dirty another pan?). Once rice is ready, add rice, onions, celery, chicken, cheese, peas, mayo, soup, pimentos and almonds (if you're using them - I don't because Marshall hates them in this casserole) in a big, big bowl. Spoon into a 9x13 inch casserole dish and top with crushed potato chips. Bake for 25 minutes or until bubbly.

This casserole also freezes really well. Just don't put the chips on until you're ready to bake. And I also put this casserole into 2 small baking dishes, which gives me a night off in the future from cooking! Just serve this with some crunchy french bread and you are WINNING!

Everybody has a beer can chicken recipe these days, so why not join in on the fun?? They even make holders these days for the chicken and the beer, which is a nice invention. When we made our first beer can chickens we just had to hope the can would stay balanced, and it always did (Barry has a funny video on our youtube channel about our balanced chickens). If you don't have a holder you can still do this recipe, just push the chicken on the can until her legs are resting on the grill or pan, depending on where you roast your bird.

Our beer can chicken is easy and no fuss. We use a lager (Miller High Life) but you can change it up and use an IPA or even pour out the beer and add wine (or use a canned wine). And if you change the flavor of beer or wine play around with spices, too. I'll give you some ideas at the bottom of the recipe.

Beer Can Chicken

serves 4-6

hands on time 10 minutes

cooking time 40-50 minutes, depending on the size of the bird and cooking technique

total time 50-60minutes

1 whole Bales Farms chicken (recommended 3.5-4.0 pounds), fully defrosted

1 can Miller High Life beer, with 3 drinks from the can (enjoy!)

2 garlic cloves, peeled and halved

1 tablespoon salt

1 tablespoon pepper

1 teaspoon onion powder

1 teaspoon paprika

1/2 teaspoon oregano

1/4 teaspoon Cayenne pepper (optional)

4 slices bacon (optional)

Crack open that beer and take 3 big drinks (if you don't want to take a drink, pour 3 tablespoons out as you don't want the liquid to overflow in your oven or grill). Then pop the garlic cloves in the beer.

Mix all spices and rub chicken generously and then pop the chicken on the beer can, resting the legs on the roasting or grill pan you are using. This will stabilize your chicken. If you'd like drape bacon slices lengthwise along chicken from the neck down.

Roast in an oven at 400 degrees for 50 minutes or grill at 350-400 degrees for 60 minutes, checking temperature often. Once the internal temperature is 165 degrees your whole chicken is ready to be enjoyed! Let it rest for 10 minutes and carve.

Note - if you're using an IPA a good combination of spices to rub would be:

1 tablespoon chili powder

1 tablespoon paprika

1 tablespoon brown sugar

1 teaspoon cumin

1 teaspoon salt

1 teaspoon pepper

If you're using white wine the above listing of spices with the Miller High Life work well. I don't recommend using a red wine with the chicken. It's too overpowering and tends to dry out the chicken in my opinion.

In both the wine and IPA options I recommend also using the method of drinking a few drinks from the can and adding garlic cloves to the can.

My White Barbecue Sauce

1 cup JFG or Blue Plate Mayo

1 tablespoon Creole mustard

1-2 teaspoons horseradish (depending on how spicy you like it)

1 garlic clove, Minced

1 teaspoon Midland Ghost White Pepper sauce

Pinch pepper, salt and sugar

Place all the ingredients in a mason jar and shake. This sauce is great with chicken, pork, seafood and steaks.

Braising meat sounds daunting sometimes but it's really pretty simple. It just takes 2 steps - searing at high dry heat and then cooking low and slow in liquid until fork tender. It's an excellent way to prepare more economic (ie, budget friendly) cuts of beef. If you don't care for these cuts of beef they turn out pretty dry and tough but braising it will give you a tender product.

This is a winter staple on the farm and it uses a CROCK POT. I don't use a crock pot very often but this recipe uses it well. It's a good way to prepare a large cut of beef and have meals all week. Please don't be shocked by the time stamp on this recipe, it's very simple. But because it cooks in the crock pot it takes some time but is hands off.

Mexican Beef Roast (aka Beef Barbacoa)

serves 8-10

hands on time 25 minutes

cooking time 4-5 hours

total time 5 hours 25 minutes

2 tablespoons olive oil

1 (3.5 pound) Bales Farms chuck or arm roast

1 teaspoon salt

1 teaspoon pepper

1 bottle Corona or Corona Light, 1 bottle of Coke or beef stock

2-3 chipotle peppers in adobo sauce

1 teaspoon adobo sauce

1 teaspoon cumin

1 teaspoon chili powder

In a large cast iron skillet heat olive oil on medium to high heat. Salt and pepper the beef roast and sear in the hot skillet for 3 minutes per side.

In a large crock pot dump liquid, chipotle peppers, adobo sauce, cumin and chili powder. Stir together and once beef is seared on both sides, add to crock pot. Cook on high for 4 hours or low for 8 hours.

Once the beef roast is fork tender remove and shred. Serve on nachos, tacos, salads, sandwiches and/or soup.

Y'all know my love for Asian food. It's tried and true. This recipe started in college when I was in Physical Therapy school and a friend gave me the golden rule for fried rice - use leftover rice that's been in the fridge for a day. Since then I've played around with it and changed it up through the years but this is my go-to recipe now. As with all good recipes the ingredient list is long and may look a bit daunting but I promise it comes together quickly. In fact I can get this from stove top to table in 15 minutes.

Here's a tip - anytime you cook rice for dinner cook extra and put it in the fridge. That way when you're running around like a chicken with it's head cut off during the week you have it and can throw this yummy dish together in no time flat.

Vegetable Fried Rice

serves 4-8

hands on time 5 minutes

cooking time 10 minutes

total time 15 minutes

4 cups cooked and chilled rice

sesame oil

1/2 red onion, diced

1 carrot, chopped

1/2 red bell pepper, chopped

1 cup frozen peas

1 egg, scrambled lightly

dressing:

3 tablespoons low sodium soy sauce

1 tablespoon oyster sauce

1 teaspoon sesame oil

toppings:

fresh cilantro, chopped

green onions, chopped

sesame seeds

Cook rice a day or so in advance and store in the fridge to allow it to dry out.

Pour sesame oil in a large skillet on medium heat. Add red onion, carrot and bell pepper and sauté 5 minutes or until fragrant and tender. Add peas and sauté for an additional 1 minute. Add rice and sauté until warm (1-2 minutes). Add dressing and stir to combine. Divide evenly into serving bowls and top as desired. Serve with soy sauce and sweet and sour sauce available.

Note - if you want meat in this fried rice, cooked Bales Farms chicken (from the buttermilk chicken recipe) works well as does pork belly and/or ground beef.

Also feel free to add more veggies - shredded cabbage in awesome in this rice!

If you have time and energy homemade pizza is the best ending to any day. Sitting outside on the porch, watching the sun set and listening to the sounds of life.... It makes everything better. And add a nice glass of wine to that scenario - you just can't beat it!

We live too far out for delivery and pizza brought in is always soggy and lukewarm so I've made pizza our whole married life. I used to always get the pre-made, packaged pizza crust but a few years ago I got serious about adulting and figured I better get bread down. So now I make my own crust and love it. It's fairly simple and straight forward but it does take a little time and effort. If you don't have it in you to make your own dough, no worries! The food police won't come knocking at your door and I bet your family won't even notice the difference. If that's your route, just skip the part about the dough and move on to the next step.

Sausage and Veggie Pizza

serves 2-4

hands on time 15-60 minutes

cooking time 5-10 minutes

total time 25-70 minutes

Pizza Crust:

1 cup warm water

2 teaspoons active dry yeast (1 packet)

1 1/4 teaspoons sugar

1 1/2 tablespoons olive oil

3 cups bread flour

1 1/4 teaspoon salt

In a small bowl mix yeast and sugar. Pour warm water on top and stir to combine. Place in an oven on proof or a warm space for 5 minutes or until bubbly. Stir in olive oil.

Mix flour and salt in a large bowl. Make a well in the middle and pour the liquid in to mix. Once mixed remove the dough from the bowl and onto your counter. Knead the dough for 5 minutes or until it is smooth and elastic (do not rush through this step! This is what makes the crust amazing.) Form the dough into a ball.

Drizzle olive oil on the inside of a bowl and place the dough ball in the bowl. Cover with a clean dish towel and let rise in a warm, dry place for 45 minutes, or until it's doubled in size (I put it back in the oven on bread proof).

Remove the bowl and punch the dough. On a work surface coated with flour work the dough into balls for pizza. I usually make 3 pizzas with my dough. If you have the oven or grill space and a pan big enough to make one huge pizza, go for it! I generally make 3 pizzas which gives us one for Marshall (cheese, sausage and spinach), one for Barry and me (sausage and all the veggies) and one crust for the freezer to use later, which makes the next time easy peasy.

Separate the dough and form into balls. Place each ball on a tray and cover to let rise a second time for 10-15 minutes, while you get your toppings ready. Then form each dough ball into a pizza shape, roll it out as thin as you like (I do mine pretty thin!) and use fork to punch the dough randomly.

Marinara sauce:

1 (28 oz) can crushed Roma tomatoes

2 garlic cloves, minced

handful fresh basil leaves, cut into strips

salt to taste

1/2 teaspoon crushed red pepper (optional)

Mix all above together. Heat in a saucepan until warm.

Toppings:

1 pound Bales Farms sweet or hot Italian sausage

1/2 red onion, sliced

1 cup fresh mushrooms, sliced

1/2 cup fresh spinach leaves

1/2 cup red bell pepper, sliced

Mozzarella cheese, shredded

To assemble pizzas, take formed pizza dough and brush generously with marinara sauce. Top with sausage, onion, mushrooms, red bell pepper and spinach then sprinkle generously with cheese. Bake in a preheated 500 degree oven for 10 minutes or on a big green egg at 450 degrees for 8-12 minutes, until the cheese is bubbly and the crust is golden brown.

My friend Laura Alley shared her meatloaf recipe with me at least 2 decades ago. Hers was the first meatloaf I ever ate and enjoyed. She generously shared it with me and, though I've changed it a bit, I still think of her and thank her every time I make it. This meatloaf recipe will give you a huge loaf pan, an 8x8 inch pan or 2 small loaf pans. I make my meatloaf in 2 small loaf pans and fix one and then freeze one. It freezes beautifully and then you have dinner ready to go when you're running around during the week.

A Different take but still kind of Laura Alley's Meatloaf

serves 8-10

hands on time 15 minutes

cooking time 60 minutes

total time 75 minutes

1 1/2 cup breadcrumbs

3/4 cup milk

2 pounds Bales Farms ground beef, defrosted

salt and pepper to your taste

2 Bales Farms eggs, lightly beaten

1 red onion, diced

1 green pepper, diced

Sauce (you'll wind up making 2 batches of this):

1/4 cup ketchup

3 tablespoons brown sugar

2 tablespoons mustard

Preheat oven to 350 degrees.

In a large bowl mix all ingredients with a large wooden spoon or your hands. In a small bowl mix the sauce and add that to your meatloaf and mix in. Spoon into a baking tray of your choice (see above) and bake for 50 minutes. Take meatloaf out of the oven and drizzle another batch of sauce on the top. Return to bake for 10 minutes.

If you are choosing to bake in smaller containers and freeze one container, defrost and bake as above, remembering to drizzle a batch of sauce at the end of baking.

Note - so, yes, if you bake in smaller batches you'll end up having more sauce in the meatloaf, which is always a welcome event in my home.

My old boss, Greg Cross, makes fun of my love of casseroles, but, really, is there anything better on a cold, wintry night than sitting down to a one-dish meal? I serve this with a salad and crusty bread and call it done. This casserole is so lovely. It's delish when you make it but it freezes really, really well. Sometimes I half the recipe and serve one and freeze one, just in smaller containers. So, bring back the 80s I say! Let's dive in with the casseroles.

Poppy Seed Chicken Casserole

serves 6-8

hands on time 10 minutes

cooking time 20-25 minutes

total time 30-35 minutes

4 cups cooked Bales Farms chicken, pulled from bones

1 (10-3/4 oz) can cream of chicken soup

16 oz sour cream

6 oz block sharp cheddar cheese, grated

2 stalks celery, diced

1/2 cup fresh mushrooms, sliced

1/8 cup poppy seeds

1 sleeve buttery crackers (I use Ritz)

1/4 cup melted butter

salt and pepper to taste

Preheat the oven to 400 degrees. Combine first 7 ingredients in a large bowl and then spoon into an 8x8 inch baking dish or large cast iron skillet. Top with crushed crackers and drizzle butter on top. Bake, uncovered, for 20-25 minutes. Season as needed with salt and pepper.

If you freeze this casserole do so before you add the crackers and butter. Add those right before you bake - you want a glorious, golden, crunchy heaven, not soggy mess.

I've already confessed my love of Asian dishes and ramen noodle bowls are no exception. Somehow I missed the crappy ramen college experience, probably because I was in the cafeteria eating chicken tenders and fries (which explained my weight gain in college!). I did grow to love noodle bowls as an adult, however, and we eat them weekly. We all love them! And they come together so quickly. They're perfect for a weeknight meal, especially since the noodles cook in 3 minutes and you use leftover meat from another night (or go vegetarian - it's completely up to you!).

In this recipe I call for pork belly. We offer pork belly here on the farm and you can use it for so many dishes. You can even make your own BACON. Yep, you read that right. You. Can. Make. Your. Own. Bacon. It's easy! It takes 3 ingredients and a little fridge space, but that's it! So try it. You'll never need to buy it from the store again.

Pork Belly Ramen Noodle Bowls

serves 4

hands on time 15 minutes

cooking time 10 minutes

total time 20-25 minutes

1 cup cubed and cooked Bales Farms Pork belly

1 tablespoon sesame oil

1 red onion, sliced thinly

1 carrot, sliced thinly

1 red bell pepper, sliced thinly

1 cup spinach leaves

3 bricks ramen noodles

dressing:

1/2 cup reduced sodium soy sauce

1/4 cup brown sugar

1/8 cup sesame oil

1 teaspoon chili garlic sauce

toasted sesame seeds to your preference

1/4 teaspoon ginger

1 tablespoon fresh cilantro leaves

garnishes:

fresh cilantro leaves

toasted sesame seeds

Bring water to boil in a large dutch oven. Once boiling add ramen noodle bricks and boil for 3 minutes. Remove from heat, drain and rinse with cold water.

Drizzle into a large skillet or wok the sesame oil. Add red onion, carrot and bell pepper and sauté for 3-5 minutes until fragrant. While this is sautéing make your dressing by adding all ingredients into a jar and shaking. Now add the spinach leaves and pork belly. Combine to heat and pour dressing in. Add noodles and stir to combine.

Pour noodles into serving bowls and top with garnish.

I've seen these bowls (and bowls like them) referred to as Buddha Bowls and grain bowls. I refer to them as quinoa bowls because I'm pretty unoriginal and I don't have much of an imagination. I pretty much call things as I see them, so these have always been referred to as quinoa bowls here. There's simple, versatile, nutritious and come together quickly for a weeknight meal. Notice the adjectives I used to describe this meal. I wouldn't necessarily serve this for a New Year's Eve bash but it does provide your family with much needed nutrition. I will say, though, that Barry and Marshall eat them and sometimes Barry even requests them. . . .

Quinoa Bowls

serves 4
hands on time 15 minutes
cooking time 30 minutes
total time 45 minutes

1 cup quinoa
2 cups water or bone broth
olive oil
1 cup red onion, diced
1 carrot, sliced thin
1 cup red bell pepper
1 garlic clove, minced
1 cup fresh mushrooms, sliced
1 cup Swiss chard, kale or spinach, sliced
1 teaspoon good quality balsamic vinegar
salt and pepper to taste
Options for toppings:
parmesan cheese, grated
Midland Ghost White Pepper sauce

In a small saucepan combine quinoa and water. Bring to a boil and then turn off heat, cover tightly and let cook for 15 minutes. Fluff with a fork.

In a large skillet over low to medium heat combine the olive oil, onion, carrot, red bell pepper and garlic. Sauté for 3-5 minutes until tender. Add mushrooms and sauté for an additional minute. Add greens and let wilt, approximately 1 minute. Drizzle balsamic vinegar and stir to combine. Season with salt and pepper.

In serving bowls, divide the quinoa evenly and top with a generous portion of veggies. If you desire top with cheese and white pepper sauce.

Note - cooked and diced chicken leftover from your buttermilk chicken works really well in this recipe if you want to add a meat option.

Everybody's got their own sloppy joe recipe, or at least instructions on the can, right? I used to never make sloppy joes but one night, in desperation, made some and the boys LOVED them. They ask for them by name now. Sloppy Joes are such a comfort food! And they can be so yummy and nutritious. Seriously. I know the canned stuff isn't healthy but when you do them yourself they're pretty good! They've got a good amount of veggies and grass fed beef. So try them sometime. The list of ingredients is long but these babies come together so so quickly. I can make a round of these and the guys eat them happily for many meals. Which means I'm happy for many meals.

In full discloser about the nutritious aspect. . . . Eaten by itself or even on homemade buns they are pretty nutritious. Eaten on tater tots covered in cheese? Umm. Delicious. (Not every meal can be healthy folks. Smothered tater tots are AMAZING.)

Sloppy Joes

serves 6-8

hands on time 10 minutes

cooking time 20 minutes

total time 30 minutes

1 pound Bales Farms ground beef

1 tablespoon olive oil

1/2 red onion, diced

1/2 green bell pepper, diced

1-2 garlic cloves, minced

1 cup tomato sauce

1 tablespoon brown sugar

1 teaspoon yellow mustard

1 teaspoon Worcestershire sauce

1/2 teaspoon kosher salt

1/2 teaspoon pepper

Saute ground beef until no longer pink in a large skillet over medium heat. Remove with a slotted spoon and reserve on a plate. Pour olive oil in skillet on medium heat and add onion, bell pepper and garlic. Sauté until tender, about 5 minutes. Add all other ingredients except beef and mix to combine. Simmer for 3-5 minutes and add 1/4 cup water if it's too thick to your taste (that's up to you). Add beef back to pan and stir to combine.

Toast hamburger buns and put sloppy joes on buns or serve without bread. Also, as stated above, these are amazing on tater tots or homemade potato chips covered in cheese. . . The good news is this recipe will make enough for you to try both ways.

Spatchcock chicken. No that's not a joke and I don't think it's a typo. But it IS about the best and easiest chicken you can make in an hour (buttermilk chicken aside). This is a great second step recipe for roasted chicken. I say second step because it does require some handling of the chicken itself and some tools and know how, so I teach this recipe as a "second step" - it's a tad bit more involved just because you have to actually touch the chicken, which some folks are against in this modern world.

So why spatchcock? Well it decreases the overall time to roast a chicken by a hair and it allows the chicken to roast evenly. You'll need a good pair of poultry shears or super duty scissors, alternatively you can use a very sharp knife. Other than that it's a breeze!

Spatchcock Chicken

serves 4-6

hands on time 15 minutes

cooking time 40-45minutes

total time 65-75 minutes

1 whole Bales Farms chicken (3.5-4.5 pounds recommended), fully defrosted

6 tablespoons butter

1 teaspoon salt

1 teaspoon pepper

1/2 teaspoon paprika

1/4 teaspoon garlic powder

1/4 teaspoon onion powder

Preheat oven to 400 degrees.

Take your chicken and place on a cutting board spine up (breast down). Using your poultry shears or scissors cut alongside the backbone and remove from chicken. Turn the chicken over to breast side up and open the chicken as if you are opening a book. Then press down on the chicken to flatten until you hear a pop. That is the breast bone breaking and you should have a very flat chicken. Now you have spatchcocked your very own chicken! Wasn't that easy?

In a large skillet over medium heat, melt the butter. Add the chicken, breast side down and sear for 2-3 minutes to a light golden color. While the chicken is searing, mix the herbs in a small bowl and sprinkle 1/2 over the chicken. Turn the chicken over and repeat the process.

Remove the skillet from the stove top and roast in the oven for 40-45 minutes, depending on the size of the chicken. You'll want to internal temperature to be 165 degrees for safety. Let the chicken rest for 10 minutes and then carve and enjoy!

Note - After I cut out the backbone I put it in the skillet beside the chicken to cook and roast. I take it out after 20 minutes as it doesn't need to roast as long as the whole chicken. Then I save the backbone along with the other bones of the chicken to make bone broth.

Remember to put the chicken in legs first as the oven is hottest in the back AND the darker meat can take heat better than white meat of the breasts. It also takes just a smidge longer to cook the dark meat which will be accomplished by being in the hottest part of the oven.

I LOVE Asian noodle dishes. I got hooked on them in college with my roommate and have loved them since then. We have Asian-inspired meals at least twice a week. They're so versatile, simple, nutritious and come together quickly. I know sometimes the ingredient list is daunting but hang in there and give it a go! Once you try one dish and see how easy and quick it is you'll be coming back for more and branching out. Super great for a weeknight meal, or weekend dinner party!

Spicy Noodles with Chicken and Chilis

serves 4

hands on time 15 minutes

cooking time 15 minutes

total time 30 minutes

1-2 cups Roasted chicken from Bales Farms whole chicken (mix white and dark meat is perfect!), diced

1 garlic clove, sliced thin

1/4 cup red onions, sliced thin

2 carrots, sliced thin

1/2 red bell pepper, sliced thin

1/2 cup broccoli, sliced thin

1 cup fresh spinach leaves

8 oz Wide rice noodles (linguine is fine if you don't have rice noodles)

Dressing:

2 teaspoons cornstarch mixed in 2 tablespoons water

1/4 cup honey

3 tablespoons low sodium soy sauce

1 tablespoon sesame oil

1-2 teaspoons Chili garlic sauce

Garnishes: fresh cilantro, toasted sesame seeds and green onions

Use more soy sauce if needed

Prepare the noodles per package. Once the water is heating in the pot for the noodles start on the rest of dinner.

In a bowl place chicken (or protein of choice) and honey, soy sauce, chili garlic sauce and sesame oil. Mix and let sit for a few minutes. Now your noodles are probably ready to go in the boiling water. They should take about 8 minutes.

Once the chicken is marinading and noodles are in the water, heat a tablespoon or so of sesame oil in a large skillet. Then add garlic and onion and let sauté for about 1-2 minutes. Next add the rest of the veggies. Let them sauté for 3-4 minutes and add the chicken and marinade. Let that sauté for a minute or so and add the cornstarch and water slurry, which will thicken the sauce.

Your noodles should be ready to go at this point so crank up the heat and ladle the noodles in the skillet. Stir those just until combined and then divide into bowls. Top with the garnishes and you're ready to eat!

This recipe is a winner any night of the week, weekend lunch or any other occasion you have leftover meat and sauce and want to whip up a delicious meal in less than 15 minutes.

When I make pizza or pasta and use sweet Italian sausage and marinara sauce I make double and then have these sandwiches on standby during the week, usually after a busy day on the farm or when we have deliveries and I have about a nanosecond to get supper prepared. By cooking a little more meat and sauce one time you'll have multiple meals you can fix quickly and save your weeknight.

Sweet Italian Subs

serves 4

hands on time 5 minutes

cooking time 5 minutes

total time 10-15 minutes

1/2 - 1 pound Sweet Italian sausage, cooked and warm

1/2 - 3/4 cup marinara sauce (leftover from pizza or pasta)

mozzarella cheese, grated or sliced

baguette bread, sliced in half and 3-4 inches in length

Toast baguettes and top with sweet Italian sausage, marinara sauce and mozzarella cheese. Return to oven to heat and melt the cheese. Serve warm.

Sides

Okay I know you almost skipped right over this because it said asparagus, didn't you? But DON'T skip this recipe!!! This recipe will save the day when you're hosting Thanksgiving! I promise. It comes together in no time. It takes about 8 minutes to roast the asparagus and less than 5 minutes for the sauce. And it's AMAZING. Truly. I have shared this recipe probably 1,589 times (okay, that could be a slight exaggeration) and everybody loves it! It's a no fail and win-win. Truly. Please promise me this Thanksgiving you'll give this a whirl. Everyone will thank you for trying it. Even your mother-in-law, who, let's face it, is a little judgmental about your cooking.

Balsamic-Browned Butter Asparagus

serves 6

hands on time 5 minutes

cooking time 8 minutes

total time less than 15 minutes

1 pound fresh asparagus, ends snapped off for freshness

1/2 teaspoon Kosher salt

1 tablespoon butter

1 teaspoon low sodium soy sauce

1/2 teaspoon balsamic vinegar (the better quality, the happier the outcome)

Preheat oven to 425 degrees. Place asparagus on a roasting pan and sprinkle lightly with salt. Roast for 8 minutes.

While the asparagus is roasting, in a small saucepan melt the butter. Do not stir. Only shake the pan. Once the butter is melted and golden brown add the soy sauce and balsamic vinegar, shaking the pan without stirring. Remove the asparagus from the oven and place in a long serving dish and drizzle with balsamic butter dressing.

Sit back and watch as everyone discovers you are the GENIUS they always underestimated. And then go get your crown to wear the rest of the day.

Confession time - I've never had boxed mac-n-cheese.

Baked Mac-n-Cheese

serves 4-8

hands on time 15 minutes

cooking time 15-20 minutes

total time 30-35 minutes

2 tablespoons butter

2 tablespoons all purpose flour

2 cups whole milk

1 teaspoon salt and pepper (or more to your taste preference)

1/4 teaspoon paprika

1/4 teaspoon Cayenne pepper

8 oz block cheese, grated (can be Cheddar, white cheddar, colby, Monterrey Jack or a combination of those)

8 oz elbow pasta, jumbo elbow or short pasta of your choice

Prepare pasta per directions on box.

Preheat oven to 400 degrees.

Make a roux by melting the butter and adding flour, whisking constantly. Once the roux is golden, add the milk and stir until thickened, about 5 minutes (you know its thick enough when it coats the back of a spoon). Remove from heat and stir in 2/3 of the cheese and seasonings.

Dump pasta into a small baking dish and stir in cheese sauce. Cover with the remaining cheese and bake for 15-20 minutes, until bubbly.

Note - you don't have to bake the mac-n-cheese if you're in a hurry. But it really is delish. Once baked it will be thicker than if you eat if straight off the stove top. I've found most adults enjoy it baked while kids like the thinner, off the stove variation better.

Here we are - CORNBREAD. Like biscuits, cornbread is a staple on the farm. Barry and Marshall both declare cornbread to be their favorite food on the planet.

I don't know why in the world so many people get the cornbread muffin mix in a bag. Cornbread is so easy to make! You just dump and stir. It's a mystery much like the fear of a whole chicken. So here is my super easy recipe for cornbread. You can't mess this up and you can whip it up in less than 45 minutes total time.

Cornbread

serves 4-6

hands on time 5 minutes

cooking time 30 minutes

total time 35 minutes

1/4 cup vegetable oil

1 egg (preferably Bales Farms)

1/8 cup vegetable oil

pinch sugar

1 cup cornmeal

1 cup whole milk or buttermilk

Preheat oven to 425 degrees. Pour 1/4 cup oil in an 8 inch cast iron skillet and put in the oven to heat up.

Mix egg, oil, sugar, milk and cornmeal in a bowl and stir to combine.

Once the skillet is sizzling hot remove from the oven and pour the mix into the skillet. Bake for 25-30 minutes, until beautifully golden brown. Flip out of the skillet and serve.

Note - I always, ALWAYS put my skillet in another pan (I use a 9 inch pie pan) in case it overflows the skillet. Wanna guess why???

Oh my love for Brussels sprouts! Like cabbage I love little Brussels sprouts in almost any form except canned. Those are from the Devil. Do not use them ever.

This is a great recipe because you can totally shred the sprouts and people who claim they hate Brussels sprouts will eat every bite and declare them brilliant, and then you can say, "really? I thought you hated Brussels sprouts?". I have served these to so many people who thought they hated the sprouts but had just been served the devil sprouts in a can covered with Velveta when they were kids. (The 80s were hard, people)

Sauteed Brussels Sprouts

serves 4-6

hands on time 10-15 minutes

cooking time 15 minutes

total time 30 minutes

1-2 slices bacon, cooked and crumbled, reserving grease

1/2 cup red onion, diced

1 pound Brussels sprouts, quartered (shredded for those who think they hate them and need to be fooled into eating them)

salt and pepper to taste

In a large skillet over medium heat, cook the bacon and remove and crumble. In the same skillet sauté the onions in the bacon grease until tender, about 3 minutes. Add the Brussels Sprouts and sauté for 5-7 minutes, then add the bacon and continue to cook for 1-2 minutes. You want the sprouts to be crisp, not soggy! Taste and add salt and pepper if needed.

Note - the pickled red onions are fabulous on this dish for a dinner party. They add a touch of elegance, crispness and beautiful color. Also a sprinkle of high quality feta or parmesan cheese would go well on this dish.

Another Note - I'm serious about shredding Brussels sprouts if you're feeding picky folks. Just put them in a food processor and pulse until mostly shredded. They'll look more like cabbage leaves at that point and everyone will realize how much they love them! Also refer to the asparagus recipe and note on the crown. You'll need your crown after this recipe, too, because everyone will instantly realize you are the QUEEN OF THE KITCHEN.

I once heard Michael Pollen say in an interview that you can have all the junk food you want as you prepare it yourself.

Game. On.

The Best French Fries I've Ever Eaten

serves 4-6

hands on time 30 minutes or so

cooking time 15 minutes per batch

total time 45 minutes

1 potato per person, cut in strips and immersed in cold water, rinsed twice

oil for frying

salt

The hardest part of this recipe is cutting of the potatoes. I allow 1 potato per person and I leave the skin on. I cut them long and thin, approximately 1/4 inch wide. I soak them in ice cold water for at least 30 minutes, rinsing twice.

Heat your deep fryer with oil to 300 degrees. Working in batches, fry the potatoes for 4-6 minutes each and remove to place on a roasting pan lined with paper towels to soak up the oil. Fry all the potatoes that way.

Once the first fry time is complete, heat the oil to 375-400 degrees. Fry to potatoes a second time for 4-5 minutes or until golden brown. Remove and sprinkle with salt of your choice (I use popcorn salt because it is super fine). Serve hot!

My guys love potatoes and we have them a lot. They love them every way - baked, fried, sauteed and roasted. This is my version of roasted potatoes.

Roasted Potatoes

serves 4

hands on time 10 minutes

cooking time 40 minutes

total time 50 minutes

1 potato per person, cut into 2 inch squares

good quality olive oil

salt and pepper to taste

1 2 inch stalk rosemary, leaves removed

Preheat oven to 400 degrees. Place large cast iron skillet with oil to coat the bottom in the oven to heat up.

Cut your potatoes and add to a large pot to boil while the oven heats up. Once the oven reaches temperature remove the potatoes from the water and drain. Place in the hot skillet and season with salt, pepper and fresh rosemary. Roast for 30-40 minutes until crisp tender.

Note - you can skip the boil on the potatoes but I feel it does make them fluffier on the inside and just more enjoyable.

Kids hate cole slaw. I think it's a rule of kid-dom. Or else it's because tired moms and grandmothers get that jar coleslaw dressing that tastes like vomit (sorry, it does) and smells even worse and pour it on the bagged slaw mix that has been sitting in the grocery store for weeks at a time and delivered from some foreign country. . . .

If we could just give people good slaw!!! Good cole slaw is worth gold to me. I love it and will defend it up to my death bed (okay, maybe that's a little dramatic but I do love cabbage in any form and especially GOOD cole slaw). I absolutely hate crap slaw. Hate it! Especially because yummy cole slaw is so easy to make!!! It literally takes less than 5 minutes and then you put it in the fridge. Come on. You can't get easier than that. And the difference??? HUGE.

Cole Slaw

serves 6-8

hands on time 5-10 minutes at most

cooking time 0

total time 5-15 minutes

1/2 cup mayonnaise

1 teaspoon sugar

1/2 teaspoon salt

1 carrot, shredded

1/4 head cabbage, shredded

Mix the mayo, sugar and salt in a big bowl.

In a food processor shred the carrot and cabbage, then add to the dressing. Put in the fridge for 1-2 hours and serve. Add more salt if needed.

Okay, how hard was that???? Please, Please, PLEASE make this slaw!!!!!!!!!!

Note - there is a long and heated debate about mayonnaises. I don't make my own because I'm not Julia Child so I buy either Blue Plate out of Louisiana or JFG out of Knoxville, Tennessee. We're pretty big on mayo around here (see almost every one of my recipes). We don't eat it by itself or anything but I do use it in a lot of baking and cooking. Those are my 2 go-to mayos. And I thank my friend, Erin Turner, for my love of Blue Plate.

I learned of this recipe when teaching Marshall world geography. This was an idea from China, where they grow lots and lots of cabbage (they are obviously near and dear to my heart). I've used this recipe so many times. I love it! It's so simple to make and goes with just about everything. Plus cabbage is such a budget-friendly choice at the grocery store. . .

Sauteed Cabbage

serves 4-6

hands on time less than 5 minutes

cooking time 10 minutes

total time 15 minutes

1/2 head cabbage, thinly sliced

sesame oil or olive oil

salt to taste

soy sauce (optional)

Heat the oil in a large skillet over medium heat. Add the cabbage and sauté for 4-5 minutes, until crisp-tender. Add the soy sauce if desired and season with salt. Serve as a side dish on it's own or incorporate with other veggies in a stir fry.

Note- I alway use reduced-sodium soy sauce.

Like Brussels sprouts I think Swiss chard gets a bad wrap. It's so good! And so easy to prepare (and grow in your garden if you can keep the deer away). This side dish is simple and delish. It goes with everything and sometimes I eat it by itself as my whole meal... Just wait and see!

Sauteed Swiss Chard

serves 4-6

hands on time 10 minutes

cooking time 10 minutes

total time 20 minutes

1 bunch of Swiss chard, chopped and placed in 2 piles - one for stems and one for leaves

1/2 cup red onion, diced

1 slice bacon, cooked, crumbled (if desired)

salt and pepper

Splash of good quality balsamic vinegar (optional)

If you're using bacon, cook it in a large skillet and remove when done, saving the grease. Add the onion and sauté for 1 minute, then all the stem portions of the Swiss chard and sauté for 3-4 minutes. Next add the leaves of the chard and sauté for 1 minute or so, letting the leaves wilt. Add the bacon and stir to combine, adding salt and pepper to season. If you're using balsamic vinegar this is the time to add it. Stir and serve!

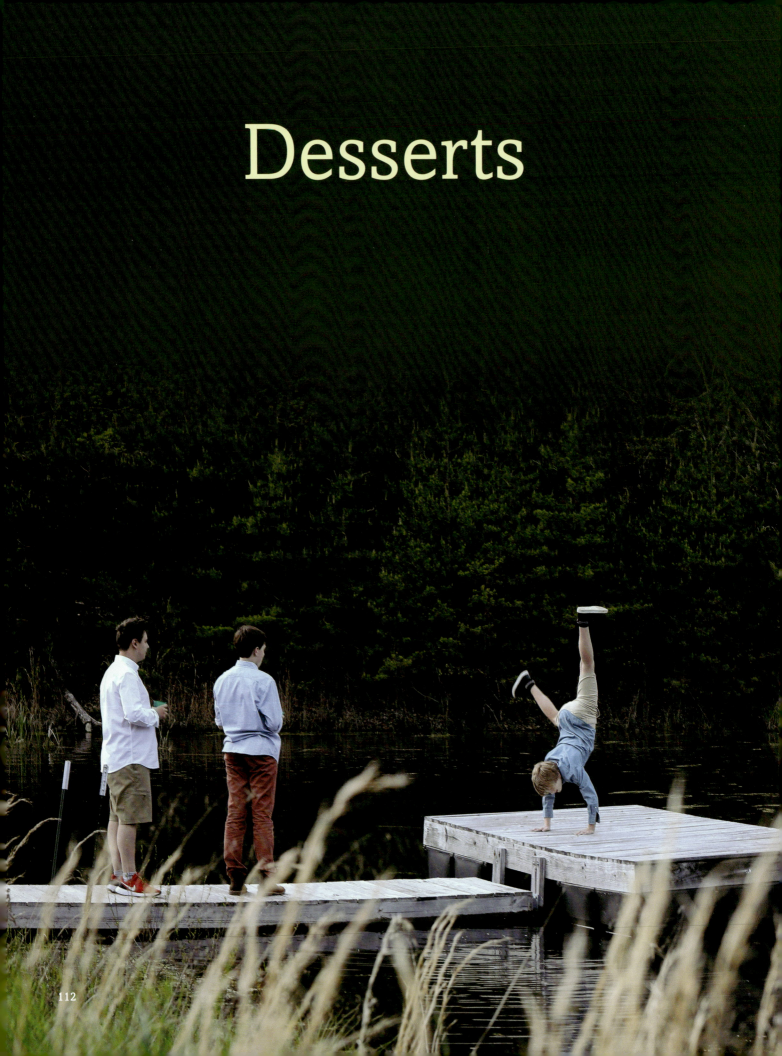

Desserts

Some say this is a fruit crisp and some say it's a cobbler but however you say it, it's delicious.

You can use about any fruit in this cobbler recipe. I use whatever's in season. Sometimes it's peach, sometimes it's cherry and sometimes it's strawberries, blueberries or blackberries (and when I don't have 4 cups of one variety I use them all). Try it with different fruit and see what you like best. It's super easy and takes absolutely no time to whip up. Just please serve it with good vanilla ice cream. It'll be a hit!

Fruit Cobbler

serves 8-10

hands on time 10 minutes

cooking time 35 minutes

total time 45 minutes

4 cups fruit of choice (I use whatever's fresh - sliced peaches, cherries, blueberries, strawberries, blackberries, or a combination)

1 Bales Farms egg, beaten

1 cup all purpose flour

1 cup sugar

4 tablespoons butter, melted

Preheat oven to 400 degrees. In a 10 inch cast iron skillet add fruit.

In a large bowl combine egg, flour and sugar and mix to combine. It should be crumbly and dry. Pour it on top of the fruit and drizzle with melted butter.

Bake for 30-35 minutes, or until bubbly and golden brown. Serve warm with ice cream.

If I'm known for a dessert, this is it. (I'm not saying I am necessarily, just that this is the most requested dessert I make.)

I've been whipping up this dessert since Barry and I got married. His great grandmother grafted her own apple trees and they grew the best apples for baking ever. Her trees aren't around and producing anymore but thankfully Barry has grafted trees for us and we look forward to apples every summer. They're super small but mighty.

Our first year of marriage Barry and I ate this pie literally every single day from June until October. With ice cream. Obviously that was 20 years and different metabolic rates ago. We don't indulge as often now, but it is fabulous!

Aliceson's Apple Pie

serves 6-8

hands on time 20 minutes

baking time 35 minutes

total time 55 minutes

2 cups all purpose flour

1 cup packed brown sugar

1 cup butter, melted

1/2 cup old fashioned oats

2/3 cup white sugar

3 tablespoons cornstarch

1 1/4 cup water

1 teaspoon vanilla

3 cups apples, peeled and diced (you can slice if you'd like, I just prefer apples bite-sized)

vanilla ice cream to top

Preheat oven to 350 degrees.

Combine flour, brown sugar, butter and oats in a large bowl. Mix well and take out 1 cup to set aside (this will become your top crust). Now press the rest into a 9 inch pie plate with your hands.

Peel and dice the apples. Drizzle vanilla in with apples and set aside.

In a small sauce pan over medium heat, combine sugar, water and cornstarch. Bring to a boil STIRRING CONSTANTLY. Do not leave for a second or you'll have a concrete mess. Once boiling this magically comes together as a syrupy beauty. Pour over the apples.

Pour apple mixture in pie plate and crumble reserved oat mixture on top. Bake for 35 minutes or until golden brown and delicious.

Please serve with vanilla ice cream! And don't even talk to me about this being fattening or not diet friendly. It's DESSERT people. Enjoy it!!!

Why is this called a Baby cake? Because that's the name I gave it, that's why! I have taken this cake to almost every friend who's had a baby in the past 10 years. It's easy and super yummy. Plus, every new mom deserves chocolate cake! And it makes the breast milk yummier (totally scientific).

This cake is easy and quick. It comes together in about an hour, give or take a few, so it is great to take to a friend with a new baby, a new family member coming to live in the house or even to take to a friend grieving. Because, new mom or not, everybody deserves to be loved on through chocolate.

Baby Cake

serves 8-12

hands on time 15 minutes

cooking time 40-50 minutes

total time 75-80 minutes

1 packaged yellow cake mix

1 (3.9 oz) package chocolate instant pudding mix

8 oz sour cream

1/2 cup vegetable oil

1/2 cup milk

4 Bales Farms eggs, beaten

1 cup semisweet chocolate chips

Preheat oven to 350 degrees. Spray or grease a regular Bundt pan. Here's a tip I use: spray the pan and then sprinkle with Rice flour. Rice flour doesn't hold water so cakes, breads and pizza crust won't stick to it. I have a bag of rice flour in the pantry at all times and it saves the day every time.

Pour cake mix, pudding, sour cream, oil, milk and eggs into a big bowl and mix until blended with a mixer. Increase speed to medium and mix for 3 minutes. Fold in chocolate chips and pour into Bundt pan.

Bake for 40-50 minutes, or until golden and a toothpick inserted in the center comes out clean. Remove from pan and let cool. Dust with powdered sugar if desired.

Okay, I admit this is a TOTAL CHEAT of a recipe. But sometimes you gotta get a cake to a friend for a celebration or to help with grief. Or sometimes you have a surprise guest coming and don't have anything on hand. This is your cake! It's a show stopper and super super simple because it uses a boxed cake mix! (You don't have to tell. Also you're not Julia Child, so it'll be okay)

This cake is total, absolute summertime and it is light, airy and delish!

Lemonade Cake with Strawberry Glaze

serves 8-12

hands on time 10 minutes

baking time 40-60 minutes, depending on pan

1 box lemon cake mix, mixed per instructions except substituting lemonade for water

glaze:

2 tablespoons strawberry jam

2 cups powdered sugar (yep, you read that right)

1 teaspoon vanilla

2-4 drops red food coloring to make a beautiful pink color (little girls go ga-ga for pink icing!)

Make and bake the lemon cake per instructions, using lemonade instead of water. This is will make an amazing lemony dessert. Remove from oven and cool completely.

Heat the strawberry jam over low to medium heat until liquid and pour into powdered sugar. Stir to combine. Add vanilla and mix. If needed to make a beautiful pink and pop against the yellow cake, add food coloring. Pour over cake.

Who doesn't love Rice Krispie treats? And I don't mean the kind you get in the packages in a vending machine - those are terrible. But homemade Rice Krispie treats are the bomb. I have a friend, Sara, who is amazing and owns a micro-flower farm as well as a cottage bakery. She reminded me of my love for Rice Krispie treats. This recipe is a wink and a nod as well as a thank you to Sara. She has her own secret ingredient that makes hers extra amazing! I won't share it, but if you happen to find yourself at a Field and Flour event, make sure you grab some of her magic. Here's mine:

Rice Krispie Treats

serves 6-8

hands on time 15 minutes

cooking time 5 minutes

4 tablespoons butter, melted

10 ounces marshmallows

6 cups Rice Krispie treats

1/2 cup sprinkles

1 cup mini marshmallows

In a large Dutch oven, melt butter over medium heat. Add marshmallows and stir until melted. Add Rice Krispie treats and stir to combine. Remove from heat and add sprinkles and mini marshmallows. Add to a 7x9 inch baking dish and refrigerate until cool. Once cool cut into squares and serve.

Note - I know most recipes call for them to be placed in a 9x13 inch dish but I love the thickness a smaller pan gives them.

Also, when you're going from the dutch oven to the pan it helps to add some non-stick cooking spray to a spatula and press into the pan.

Breakfast

Okay, I know. Everybody has their own banana bread recipe - kind of like how everybody has their own zucchini bread recipe. So you may want to just skip this recipe. If you don't have a banana bread you love, try this one. I'll go ahead and let you know I don't use nuts in my banana bread, but feel free to throw them in. I made this recipe years ago for a friend with a nut allergy and realized I much prefer it without nuts. But if you love almonds, pecans or (ugh) walnuts, throw those babies in!

Banana Bread

serves 10-12

hands on time 5 minutes

baking time 60-75 minutes

2 cups all purpose flour

1 teaspoon baking soda

1/4 teaspoon salt

1 teaspoon cinnamon

1 egg

1 cup white sugar

1/4 cup brown sugar

1/2 cup vegetable oil

2 tablespoons buttermilk or whole milk

1 teaspoon vanilla

3 bananas, mashed using a potato masher

Preheat oven to 325 degrees. Add dry ingredients to a large bowl. In a small bowl mix egg, sugars and oil, stirring to combine. Add that to the dry ingredients and stir.

In the small bowl you just emptied, mash the bananas and add the milk and vanilla. Mix into the batter and stir to combine as smoothly as possible. Pour into a regular loaf pan and bake for 60-75 minutes, until set in the center.

Note - most of the time when I make this the top of the bread gets brown before the center is baked. I tent aluminum foil over the pan and continue to bake until done.

Dutch babies! Puff pancakes! Whatever you call these creations they're lovely, beautiful, delicious, fanciful, decadent and deceptively simple to make. These come together so quickly and appeal to everyone I know. If you make these for overnight guests they will be amazed and delighted. And they will come back again and again.

I am going to share with you my two favorite Dutch Babies in this book - apple and chocolate. You can make other flavors but these are my favorites.

Apple Dutch Baby

serves 6-8

hands on time 20 minutes

cooking time 15 minutes

total time 35 minutes

4 tablespoons butter

3 apples, peeled and thinly sliced

3/4 cup whole milk

4 Bales Farms eggs

3/4 cup all purpose flour

pinch salt

1/4 teaspoon baking powder

1 tablespoon sugar

Topping:

1/2 cup sugar

1 tablespoon cinnamon

Heat your oven to 400 degrees and put a 10 inch cast iron skillet in the oven with butter to allow the butter to melt. Once the butter is melted remove and swirl the skillet to coat with the butter, even up the sides (this is what will allow the Dutch baby to rise like a souffle). Add the apple slices in a single layer.

In a large bowl mix the dry ingredients and in a smaller bowl mix the wet ingredients. Mix together and whisk until smooth. Pour the batter over the apples and sprinkle the topping on top. Bake for 15-20 minutes, or until the Dutch Baby is poofy and beautiful. Once it sits at room temperature it may deflate slightly so try to make it right before you serve it.

Note - Some people serve this Dutch Baby with syrup. I don't because, honestly, I don't think it needs it. But if you're serving someone with an extreme sweet tooth you might wanna have some warm and ready.

I don't have a lot of kitchen gadgets. I long for a stand mixture but haven't made the leap. I also have eschewed a bread mixer, instant pot and air fryer. I have a few gadgets but try to minimize my clutter in the kitchen. One of the gadgets I have that I do enjoy, though, is my waffle maker. I bought it several years ago when my nephew lived with us and made waffles almost every weekend because they were such a hit. I don't make them as often now but every time I do they are gobbled right up.

These waffles freeze well. So I make them and any leftover are wrapped in wax paper and put in a ziplock bag. To thaw I remove and place them on 2 paper towels. Once thawed, heat in the oven. They're great that way and will stay fresh in the freezer for a couple of months.

These waffles aren't healthy. They just aren't. But every once in a while everybody needs a treat. So treat yourself and those you love. And make these babies decadent!

Belgian Waffles

serves 4-6
hands on time 10 minutes
cook time 3-5 minutes per batch
total time 15 minutes

2 cups all purpose flour
4 teaspoons baking powder
1/4 teaspoon salt
1/4 cup sugar
1/2 cup vegetable oil
2 cups whole milk
1 teaspoon vanilla extract
2 Bales Farms eggs, separated

Preheat your waffle maker to the desired setting.

Add the dry ingredients in a large bowl.

In a medium bowl mix together egg yolks, milk, oil and vanilla with a whisk until smooth. Add the wet and dry ingredients and whisk again until smooth.

In a small bowl beat egg whites until stiff peaks are formed with a mixer. (You can do this by hand if you need a shoulder work out but otherwise save time and energy using your mixer.)

Gently fold the egg whites into the batter.

Spray a nonstick cooking spray on your waffle maker and add the waffle batter as directed by your maker. Cook until the maker signals the waffle is done and serve.

Note - I always serve these waffles with fruit, powdered sugar and whipped cream. I occasionally add chocolate chips and chocolate syrup if I'm serving them with strawberries. I mean, if you're going in you might as well go all in!

Chorizo hash... What can I say? It's a perfect breakfast, brunch, lunch or dinner. It's simple - everything goes in one pot! Literally the hardest part of preparing chorizo hash is the chopping of the veggies. Once this meal is started you can hop in the shower to get a jump start on the day or return messages before dinner. It's that easy. Chop, toss and run.

I prepare chorizo hash for breakfast on days I know we're going to be busy on or off the farm. Times when lunch might be late or not at all. Or times when lunch was skipped because someone decided to get out of their fence and have a field day; like I said it's great for any meal.

Chorizo Hash

serves 4-6

hands on time 15 minutes

cooking time 25 minutes

total time 40 minutes

1/2 pound Bales Farms chorizo

3 medium potatoes, diced into bite size squares

1 red onion, diced

1/2 red bell pepper, diced

1 bunch Rainbow chard, stems diced and leaves torn or

spinach or kale

4-6 eggs (2 per person)

Salt and pepper to taste

Sauté the chorizo in a skillet on medium heat until it's cooked through (no longer pink) and remove from the skillet.

Add potatoes and salt and pepper to season. Let the potatoes sauté for 10-15 minutes until they start to tenderize. You may need to add either olive oil or butter (your preference) as the chorizo is very lean. You just don't want to burn the potatoes!

Now add your onion, bell pepper and rainbow chard stems. Let sauté for 5 minutes. Add the chard leaves and chorizo and sauté for an additional 3-5 minutes.

Make a well for each egg and crack an egg in, cooking until over easy.

These cookies are so good! And they're nutritious! They are flexible and easily changed to meet the needs of your family and use whatever's in your pantry. You can't go wrong with these little treats. And telling your kids they're having cookies for breakfast? Win-win.

Breakfast Cookies

serves 10-12

hands on time 15 minutes

cooking time 15 minutes per batch

total time 30-45 minutes

3 ripe bananas smashed

2 tablespoons local honey

1/2 cup crunchy almond butter

1 egg

2 1/2 cups old fashioned oatmeal

1/2 cup flax seed

1 teaspoon baking powder

1-2 teaspoons cinnamon (depending on how much you like)

Fresh or dried fruit

1/3 cup milk chocolate chips

Preheat oven to 350 degrees. Mix bananas, honey, almond butter and egg together in a large bowl. In another bowl, mix oatmeal, flax seed, powder and cinnamon. Add the wet ingredients to the dry and mix. Then fold in fruit and chocolate chips. Place cookies in a teaspoon or tablespoon size blob on a baking sheet and bake for 12-15 minutes. Once cooled place in a ziplock bag or Tupperware container and keep refrigerated for up to 10 days (they won't last that long though).

Here's the second Dutch Baby and it's chocolate. This poofy pancake is so good you could serve it as a dessert but I always serve it for breakfast. It's magical on a snowy, wintry morning or in the summer with fresh berries. And for Christmas morning it's just such a treat. I can't recommend this enough! And it's just so easy. So treat yourself!

Chocolate Dutch Baby

serves 6-8

hands on time 15 minutes

cooking time 15 minutes

total time 30 minutes

3 tablespoons butter

3/4 cup whole milk

3 Bales Farms eggs

1 teaspoon vanilla extract

1/3 cup all purpose flour

1/4 cup unsweetened cocoa powder

1/4 teaspoon salt

1/4 cup sugar

Toppings:

powdered sugar

strawberries and/or strawberry jam, heated

Preheat oven to 400 degrees and place a 10 inch cast iron skillet in the oven with the butter to allow the butter to melt. Once melted swirl the skillet to coat the bottom and sides, which will allow the Dutch Baby to rise to it's glorious self.

In a large bowl combine the dry ingredients. In a small bowl combine the milk, eggs and vanilla. Add together and whisk to combine into a smooth, silky batter. Pour the batter into the skillet and bake for 15 minutes or until the Dutch baby is poofy, beautiful and set.

Remove and sprinkle with powdered sugar and berries.

Note - I pour strawberry jam in a small sauce pan and reduce over low heat. Then I drizzle that over the Dutch Baby for a fantastic breakfast!

I don't make this bread too often - it's too tempting. But when I do I serve it with cinnamon-brown sugar cream cheese and people scarf it up. It's pretty amazing and easy to make. But watch out - it's addictive for sure.

I do often take this to friends who need a bit of love - either for new babies or when someone is sick. It just makes you feel good knowing you don't have to worry about breakfast when you've got so much on your mind and heart. So love on your people.

Cinnamon Bread

serves 10-12

hands on time 10 minutes

cooking time 50 minutes

total time 60 minutes

1/2 cup butter at room temperature

1 cup sugar

2 cups all purpose flour

1 teaspoon baking soda

1 Bales Farms egg

1 cup whole milk

Cinnamon-sugar mixture:

1/2 cup sugar

1 teaspoon cinnamon

Preheat oven to 350 degrees.

In a large bowl add egg, butter and sugar and mix with a mixer at low speed until creamy.

In a medium bowl add milk, flour and soda. Add the two bowls together and mix to incorporate.

In a small bowl mix cinnamon and sugar.

In a loaf pan add 1/2 the batter, then most of the cinnamon-sugar mix. Add the rest of the batter and sprinkle the top with the cinnamon-sugar mix. Swirl with a butter knife and bake for 50 minutes, or until the batter is set.

If you want a healthier and more economic alternative to boxed cereal, this is your answer! I've been making this granola for 22 years. I make it every 2 weeks and it is always in our pantry. Everybody loves it and it is easy, versatile, nutritious and delicious. Barry eats it with milk as a cereal, Marshall eats it with fruit and yogurt as a parfait and my dad eats it as a snack dry. You can't go wrong!

Granola

serves 12-16

hands on time 10 minutes

cooking time 40 minutes

total time 50 minutes

4 cups old fashioned oats

1 cup sliced almonds

1/3 cup flax seed

1 teaspoon cinnamon

2 tablespoons vegetable oil

1/3 cup molasses

1/3 cup honey

1/3 cup water

1 teaspoon vanilla extract

Preheat oven to 325 degrees.

In a 9x13 inch baking dish combine oats, almonds, flax seed and cinnamon.

In a small sauce pan pour oil first and then molasses, honey and water (using oil in your measuring cup first will ensure your honey and molasses pour out of the measuring cup). Bring to a boil stirring constantly. Once boiling, remove from heat and stir in vanilla. Pour over oats mixture and bake for 40 minutes, stirring every 10 minutes.

Note - this granola is great with other nuts and/or dried fruit in it. I typically don't put dried fruits in because we use a lot of fresh fruit when we serve it. But dried cranberries, cherries, apricots, coconut and/or blueberries go well.

We've come to one of Barry's favorite breakfasts - Huevos del campo. Barry likes his eggs fried and a little spicy. This is his go to breakfast day in and day out. They didn't have a name until a friend named them for us, so thank you Jennifer Lawson!

Huevos del Campo

serves 1

hands on time less than 5 minutes

cooking time 5-10 minutes

total time 15 minutes

2 corn tortillas

2 Bales Farms eggs

1 to 2 tablespoon Guacamole

1 tablespoon pico de gallo

Midland Ghost pepper sauce to taste

salt and pepper to taste

fresh cilantro to garnish

In a small pan fry the eggs to your preference. In a hot oven broil the tortillas to your preference. Remove from heat and smear the guacamole on the tortillas. Cover the guac and tortilla with one egg each. Top with pico and hot sauce if desired. Add salt and pepper to taste.

I have to confess. I was terrible at biscuits for the longest time! And I really didn't think you could tell the difference between frozen biscuits and homemade (probably because mine were terrible). I'm just not a great baker. My mom is, though, and by her loving and wise counsel I've perfected the biscuit-making process and the recipe here is for NO FAIL biscuits. Promise.

No Fail Biscuits

serves 4-8

hands on time 25 minutes

cooking time 15 minutes

total time 40-50 minutes

1 stick butter, frozen

2 1/2 cups self rising flour

1 cup whole milk or buttermilk

1 tablespoon butter, melted

Preheat your oven to 475 degrees.

Using a box grater, grate the frozen butter in a large bowl. Add the flour and mix to combine. Refrigerate for 10 minutes.

Add milk to flour and stir to combine using a wooden spoon.

On the counter, sprinkle a bit of flour and dump the dough onto the counter. Knead with your hands until it's combined and forms a round ball. Using your rolling pin, roll to the thickness you desire (your biscuits will rise in the oven). I suggest about 1/2 inch thickness. Using a glass or biscuit cutter, cut out your biscuits and place in an iron skillet letting the sides barely touch. Once all biscuits are in the skillet, brush with melted butter.

Bake for 15 minutes in the oven. They'll be beautiful and delish! Relish in your biscuit expert making! You've done enough work for the entire day!!!

Well well well. Here we are. The true Southern condiment. Sausage Gravy. How many times have I made this in my marriage?? Too many to count, let me tell you. This is another recipe that isn't. It's not a recipe so much as a suggestion. A list of ingredients to help you on your way, but your way will be different than mine because everybody who eats gravy likes theirs thicker or thinner or runnier or with coffee. Every family is different and I think it's just how you were raised. I never ate gravy growing up so didn't have an opinion. I just make it now how Barry and Marshall like it and hope it's the right seasoning. I can get the consistency down, which is what you'll have to master for your own family.

Sausage Gravy

serves 4

hands on time 5-15 minutes

cooking time 5-15 minutes

total time 5-15 minutes

1 pound Bales Farms breakfast sausage, cooked and removed, saving the grease

1/4 cup all purpose flour

1/4-1/2 whole milk

pepper to taste (quite a bit)

paprika to taste (just barely)

Remove sausage from the skillet and keep warm in the oven. Save the grease. In the skillet sprinkle the flour in the grease and stir to combine, making a roux. Stir until all the flour is incorporated. Slowly pour in the milk until the consistency you want and then go a teaspoon more (it'll thicken up, don't worry). Add the pepper and paprika and continue to stir. Add milk if it's too thick and flour if it's too runny.

Who doesn't love a good pancake on a Saturday morning?

This recipe is probably older than most of us put together. I found it in an old, old recipe book my grandmother had. She even made her own maple syrup to go with it! I tried her recipe for that but we just couldn't do it so stick with some pure maple syrup our friends from upstate New York are kind enough to send to us.

I make these puppies a few times a month and freeze a couple each time. They pop out of the freezer just fine and heat up as if they were fresh that morning! So if I were you I'd do just that and pretend you've worked all morning on them.

Buttermilk (or whole milk) Pancakes

serves 4-6

hands on time 10 minutes

cooking time 5 minutes per batch

total time 20 minutes per batch

1 Bales Farms egg

1 cup all purpose flour

1 teaspoon baking powder

1/2 teaspoon baking soda

1/2 teaspoon salt

1 tablespoon sugar

1 cup buttermilk or whole milk

2 tablespoons vegetable oil

Beat egg in a big bowl and combine all ingredients. Mix until smooth.

Pour 1/4 cup batter on a hot electric skillet, griddle or iron skillet for each pancake. When the pancakes have bubbles on the surface they are ready to be flipped. Once flipped cook until heated through.

Serve with your favorite syrup and enjoy!

Note - I often do either chocolate chip or blueberry pancakes and here's how I do them. For the chocolate chip variety I add chocolate chips to the batter and then to serve I sprinkle the pancakes with powdered sugar, drizzle with chocolate syrup and top with more chocolate chips and whipped cream. For the blueberry variety I add blueberries to the batter and serve with blueberry syrup.

We are loving frittatas these days. They're so easy and versatile. They can be all veggies or all meat or all cheese. Whatever floats your boat that morning. And I love I can chop for a few minutes, throw everything in the oven and run for the shower, and come out clean to a homemade breakfast!

Veggie (maybe with bacon?) Frittata

serves 2-3

hands on time 15 minutes

cooking time 10-15 minutes

total time 25-30 minutes

1 teaspoon olive oil

1 garlic clove, minced

1/2 cup red onion, diced

1/2 cup red bell pepper, diced

1/2 cup fresh mushrooms, sliced

1 cup fresh spinach leaves

1 slice Bales Farms pork belly (optional)

3-4 Bales Farms eggs, lightly beaten

1/2 cup Monterrey or Colby cheese, grated and divided

salt and pepper to taste

Preheat oven to 400 degrees. Heat oil in an 8 inch cast iron skillet on low to medium heat. Add onion, garlic, mushrooms and bell pepper and sauté for 3-5 minutes, until the veggies begin to get tender. Add spinach and sauté for 1 minute, until the spinach starts to melt. Pour the eggs on top and season with salt and pepper. Sprinkle 1/2 the cheese on top and place in the oven for 10-15 minutes, or until the eggs are set. Remove from the oven and sprinkle with remaining cheese.

You can change up the veggies if you'd like. Or add more to the pile! You want the same cups of veggies and the number of eggs.

People and Products We Love

Okay here's a listing of businesses I adore. I love the people, their values and their sense of purpose. Here's my incomplete list as I know I'm missing some folks but these businesses are featured here in this book and that's why I'm listing them.

First of all, Tina Wilson Photography and Reclaimed Inspired Goods. This book wouldn't exist without Tina. She encouraged us and cheered for us from day one. And Tina has done our photographs since Marshall was a baby and there's no one I trust more with my memories and to document my days. Tina is one of my dearest friends and I love her wisdom, joy, encouragement, edification and our shared laughter through tears. As they say in Steel Magnolias, "laughter through tears is my favorite emotion!" Tina Wilson Photography and Reclaimed Inspired Goods can be found physically at 414 South Roan Street, Johnson City, TN 37601 and on the web at www.reclaimedinspired.com.

Field and Flour is a micro-flower farm in Greene County and if you follow us on social media you have definitely heard of them and seen their beautiful creations (and me holding half eaten cookies made by Sara). Sara and Josh Maximoff own and operate Field and Flour and are amazing humans and friends. My Rice Krispie treats are a knock off of Sara's and hers are much better than mine (I did give her credit in the recipe!). Sara is brilliant and beautiful. She is kind and encouraging. She lifts me up when I am down and she makes me a better human. They can be found at www.field-and-flour.com.

Midland Ghost produces small batch hot sauce of every variety. Physically they're out of Georgia and Barry met David years ago before they became the celebrities they are now so we have loved their sauces for years and years. They ship and your package will arrive safely and in the best packaging. I'm so happy for their success! I order from them throughout the year and send their products as gifts for Christmas each year. They are, simply, the best and can be found at www.midlandghost.com. Get the white pepper sauce. Trust me. I use it on nachos, sandwiches, in marinades and salad dressings and even on a hash in the morning.

Zi Olive is the company I recommend for all olive oils and vinegars. Don Goins is the best teacher and encourager I've ever known. He is phenomenal at what he does, which is educating you on why you need his oils and vinegars and why buying high quality ingredients matters. Before Don I used whatever I found in the grocery store. Now I drive 75 minutes to buy his products. If you're not in Gatlinburg (see previous notes on Don's amazing store) you can find him online at www.ziolive.com. He's worth it.